SUPPLY CHAIN PERFORMANCE MEASUREMENT IN TEXTILE ENTERPRISES

Frontiers of Mechanical and Industrial Engineering

SUPPLY CHAIN PERFORMANCE MEASUREMENT IN TEXTILE ENTERPRISES

Pranav C. Charkha, PhD
Santosh B. Jaju, PhD
Prasenjit Chatterjee, PhD

AAP | APPLE ACADEMIC PRESS

First edition published 2024

Apple Academic Press Inc.
1265 Goldenrod Circle, NE,
Palm Bay, FL 32905 USA

760 Laurentian Drive, Unit 19,
Burlington, ON L7N 0A4, CANADA

CRC Press
2385 NW Executive Center Drive,
Suite 320, Boca Raton FL 33431

4 Park Square, Milton Park,
Abingdon, Oxon, OX14 4RN UK

Library and Archives Canada Cataloguing in Publication

Title: Supply chain performance measurement in textile enterprises / Pranav C. Charkha, PhD, Santosh B. Jaju, PhD, Prasenjit Chatterjee, PhD.
Names: Charkha, Pranav C., author. | Jaju, Santosh B., editor. | Chatterjee, Prasenjit, 1982- editor.
Description: First edition. | Series statement: Frontiers of mechanical and industrial engineering | Includes bibliographical references and index.
Identifiers: Canadiana (print) 20230568300 | Canadiana (ebook) 20230568343 | ISBN 9781774913864 (hardcover) | ISBN 9781774913871 (softcover) | ISBN 9781003400240 (ebook)
Subjects: LCSH: Business logistics—Management. | LCSH: Textile industry—Management.
Classification: LCC HD38.5 .C43 2024 | DDC 658.5—dc23

Library of Congress Cataloging-in-Publication Data

Names: Charkha, Pranav C., author. | Jaju, Santosh B., author. | Chatterjee, Prasenjit, 1982- author.

Title: Supply chain performance measurement in textile enterprises / Pranav C. Charkha, PhD, Santosh B. Jaju, PhD, Prasenjit Chatterjee, PhD.

Description: First edition. | Palm Bay, FL : Apple Academic Press, [2024] | Series: Frontiers of mechanical and industrial engineering | Includes bibliographical references and index. | Summary: "This new volume provides detailed insight into supply chain management (SCM) and supply chain performance measurement (SCPM) in the textile industry. Supply Chain Performance Measurement in Textile Enterprises offers the information needed to understand, implement, and evaluate the supply chain performance of the textile industry and its cyclic processes. The volume will also facilitate the textile industry by offering a comprehensive set of performance sub-criteria for the evaluation of supply chain performance measurement and elaborating on the practices of supply chain performance in identified textile industries. Using the Delphi method (a process used to arrive at a group opinion or decision by surveying a panel of experts), the authors studied several companies of diverse sizes and applied an analytical hierarchy process to establish crucial performance measurement criteria and sub-criteria. Around 80 textile industry personnel were contacted, including general managers, factory managers, managers, assistant managers, executives, and supervisors involved in various function related to textile supply chains to discuss selected procurement, production, and distribution alternatives. The authors employed the multi-criteria decision-making technique, i.e. analytical hierarchy process, for executing pair-wise comparison and establishing priority weights of criteria, sub-criteria, and for estimating the contribution of supply chain cyclic processes to overall supply chain performance. The book will be important to the textile industry as it provides a comprehensive set of performance criteria (BSC perspectives) and measures (sub-criteria) and evaluates the contribution of each of the supply chain cyclic processes towards overall supply chain performance, thereby enabling managers of independent units or partners of the supply chain to improve the performance of supply chain as a whole"-- Provided by publisher.

Identifiers: LCCN 2023051054 (print) | LCCN 2023051055 (ebook) | ISBN 9781774913864 (hardback) | ISBN 9781774913871 (paperback) | ISBN 9781003400240 (ebook)

Subjects: LCSH: Textile industry--Management. | Business logistics--Management.

Classification: LCC HD9850.5 .C43 2024 (print) | LCC HD9850.5 (ebook) | DDC 677/.0282--dc23/eng/20231101

LC record available at https://lccn.loc.gov/2023051054

LC ebook record available at https://lccn.loc.gov/2023051055

ISBN: 978-1-77491-386-4 (hbk)
ISBN: 978-1-77491-387-1 (pbk)
ISBN: 978-1-00340-024-0 (ebk)

ABOUT THE FRONTIERS OF MECHANICAL AND INDUSTRIAL ENGINEERING BOOK SERIES

EDITORS-IN-CHIEF

Dr. Prasenjit Chatterjee
Department of Mechanical Engineering, MCKV Institute of Engineering, West Bengal, India

Dr. Dilbagh Panchal
Department of Industrial and Production Engineering,
Dr. B. R. Ambedkar National Institute of Technology (NIT) Jalandhar, Punjab, India

Description of the Series

Mechanical engineering applies the principles of mechanics and materials science for analysis, design, manufacturing, and maintenance of mechanical systems, whereas industrial engineering involves figuring out how to make or do things better with minimum resource consumption. Mechanical and industrial engineering utilize a deep foundational understanding of mathematics, physics, and analysis to develop machines and systems. The former specializes in thermodynamics, combustion, and electricity to make complex machines. The latter focuses on designing workflow and making production efficient.

This series reports on the cutting-edge developments to present novel innovations in the field of mechanical and industrial engineering. Volumes published in this series aim to embrace all aspects, subfields, and new challenges of mechanical and industrial engineering and to be of interest and use to all those concerned with research in various related domains of mechanical and industrial engineering disciplines. The series aims to provide comprehensive value-added resources to the body of the literature to serve as essential references for years ahead. Features of the volumes under this series explore recent trends, model extensions, developments, solutions to real-time problems, and case studies.

The coverage of the series is committed to provide a diverse yet rigorous research experience that builds on a solid knowledge of science and engineering fundamentals and application-oriented knowledge of mechanical and industrial engineering practices for efficient functioning in today's global multidisciplinary rapidly changing work environment. It includes all new theoretical and experimental findings in mechanical, industrial engineering, and allied engineering fields.

Topics and Themes:

The primary endeavor of this series is to introduce and explore contemporary research developments in a variety of rapidly growing research areas. The volumes will deal with the following topics but not limited to:

- Automotive engineering
- Aerospace technology
- Big data in mechanical / manufacturing / industrial engineering
- Computational methods for optimizing manufacturing technology, robotics, mechatronics
- Decision science applications in mechanical, manufacturing, and industrial engineering
- Design and optimization of mechanical components
- Dynamical systems, control
- Engineering design
- Engineering thermodynamics, heat, and mass transfer
- Evolutionary computation
- Finite element method (FEM) modeling/simulation
- Fluid mechanics
- Fuzzy logic and neuro-fuzzy systems for relevant engineering applications
- Green and sustainable engineering techniques for modern manufacturing
- Innovative approaches for modeling relevant applications and real-life problems
- Life cycle engineering
- Machinery and machine elements
- Manufacturing

- Mathematical concepts and applications in mechanical and industrial engineering
- Materials engineering
- Mechanical structures and stress analysis
- MEMS
- Modeling in engineering applications
- Nanotechnology and microengineering
- Non-conventional machining in modern manufacturing systems
- Precision engineering, instrumentation and measurement
- Numerical simulations
- Soft computing techniques
- Sustainability in mechanical, manufacturing, and industrial engineering
- Theoretical and applied mechanics
- Tribology and surface technology

The series aims to serve as a valuable resource for undergraduate, postgraduate, doctoral students, researchers, academicians, and industry professionals in mechanical, manufacturing, industrial engineering, product design, management, and more.

BOOKS IN THE SERIES

Optimization Methods for Engineering Problems
Editors: Dilbagh Panchal, PhD, Prasenjit Chatterjee, PhD, Mohit Tyagi, PhD, and Ravi Pratap Singh, PhD

Modern Manufacturing Systems: Trends and Developments
Editors: Rajiv Kumar Garg, PhD, Ravi Pratap Singh, PhD, Rajeev Trehan, PhD, and Ramesh Singh, PhD

Multi-Criteria Decision-Making Methods in Manufacturing Environments: Models and Applications
Shankar Chakraborty, PhD, Prasenjit Chatterjee, PhD, and Partha Protim Das, PhD

Supply Chain Performance Measurement in Textile Enterprises
Editors: Pranav G. Charkha, PhD, Santosh B. Jaju, PhD, and Prasenjit Chatterjee, PhD

Electro-Micromachining and Microfabrication: Principles and Research Advances
Editors: Sandip Kunar PhD, Golam Kibria, PhD, and Prasenjit Chatterjee PhD

Robotics and Automation in Healthcare: Advanced Applications
Editors: R. Thanigaivelan, PhD, Sanjay Singh, PhD, and Clement Christy Deepak C., PhD

Optimization of Advanced Manufacturing Processes
Editors: Sandip Kunar, PhD, Prasenjit Chatterjee, PhD, and M. Sreenivasa Reddy, PhD

ABOUT THE AUTHORS

Pranav C. Charkha, PhD
Head and Dean Academics, Pimpri Chichwwad Education Trust's Nutan Maharashtra Institute of Engineering and Technology, Maharashtra, India.

Pranav C. Charkha, PhD, is working as Head and Dean Academics at Pimpri Chichwwad Education Trust's Nutan Maharashtra Institute of Engineering and Technology, Maharashtra, India. He is also associated with the Indian Society of Technical Education, Institutions of Engineers, and Indian Institution of Industrial Engineering. His teaching interest areas are industrial engineering, computer-integrated manufacturing, manufacturing system integration, and management, supply chain management (elective course). His research areas are supply chain management, additive manufacturing, and world-class manufacturing for industry 4.0. He has published more than 30 research papers in international journals (Springer and Inderscience) and has presented papers at many international conferences. He has been awarded with three copyrights and has published two patents along with four submitted patents. He is a coordinator for the Center of Excellence (Maharashtra Region) of Visionary Learning Community of India (VLCI) for creating Visionary Leaders for Manufacturing (VLFM) (an initiative of CII and the Japanese International Cooperation Agency). He is also a facilitator for creating synergy between academic and industry, working on a project with the German Agency for International Cooperation (GIZ). He is also actively involved in various industry-linked initiatives for students.

Santosh B. Jaju, PhD
Professor in the Department of Mechanical Engineering of G. H. Raisoni College of Engineering, Maharashtra, India.

Santosh B. Jaju, PhD, is presently working as Professor in the Department of Mechanical Engineering of G. H. Raisoni College of Engineering Maharashtra, India. He has more than 22 years of teaching experience. He has to his credit more than 100 research papers published in national and international journals and conferences as well as one book and six book chapters. He has attended more than 50 symposiums, short-term courses, training sessions, workshops, etc. He serves as a as an editorial board and reviewer for several international journal and conferences. He is a certified Lead Auditor of ISO 9001: 2008 Quality Management System by IRCA UK. He has guided more than 20 postgraduate dissertations, with four PhDs awarded, and one student having submitted his thesis. He has filed six patents, and completed research projects of All India Council for Technical Education. He received a best teacher award given by the G. H. Raisoni College of Engineering for the year 2003–2004. He has also received a best case study award (nontraditional area) in the form of silver medal from the National Council of the Indian Institution of Industrial Engineering. His research area includes quality cost, TQM, SCM, service quality, lean manufacturing, productivity improvement techniques, and industrial engineering areas.

Prasenjit Chatterjee, PhD
Professor of Mechanical Engineering and Dean (Research and Consultancy), MCKV Institute of Engineering, West Bengal, India.

Prasenjit Chatterjee, PhD, is currently Professor of Mechanical Engineering and Dean (Research and Consultancy), MCKV Institute of Engineering, West Bengal, India. He has over 100 research papers published in international journals and peer-reviewed conferences. He has authored and edited more than 15 books on intelligent decision-making, supply chain management, optimization techniques, risk, and sustainability modeling. He has received numerous awards, including a Best Track Paper Award,

Outstanding Reviewer Award, Best Paper Award, Outstanding Researcher Award, and University Gold Medal. Dr. Chatterjee is the Editor-in-Chief of the *Journal of Decision Analytics and Intelligent Computing*. He has also been a guest editor of several special issues in different SCIE/Scopus/ESCI (Clarivate Analytics) indexed journals. He is the Lead Series Editor of the book series, Smart and Intelligent Computing in Engineering; Founder and Lead Series Editor of several book series, including Concise Introductions to AI and Data Science; AAP Research Notes on Optimization and Decision-Making Theories; Frontiers of Mechanical and Industrial Engineering; and River Publishers Series in Industrial Manufacturing and Systems Engineering. Dr. Chatterjee is one of the developers of two multiple-criteria decision-making methods: Measurement of Alternatives and Ranking according to COmpromise Solution (MARCOS) and Ranking of Alternatives through Functional mapping of criterion sub-intervals into a Single Interval (RAFSI).

CONTENTS

ABBREVIATIONS

ABC	activity-based costing
AHP	analytic hierarchy process
ANP	analytic network process
BPR	business process reengineering
BSC	balance scorecard
CAGR	compound annual growth rate
CBR	case-based reasoning
CFA	confirmatory factor analysis
CR	consistency ratio
CSF	critical success factor
DEA	data envelopment analysis
DMAIC	define, measure, analyze, improve, and control
DMU	decision-making units
DSM	decision structure matrix
FLC	fuzzy linguistic computing
GA	genetic algorithm
HRD	human resource department
IIP	index of industrial production
IT	information technology
JIT	just in time
KM	knowledge management
KPIs	key performance indicators
MAUT	multi-attribute utility theory
OEM	original equipment manufacturer
PBC	performance-based costing
PM	performance measurement
PMS	performance measurement system
QE	quality engineering
QFD	quality functional deployment
QR	quick response
QRPMS	quantitative relationship performance measurement system
ROI	return on investment

SC	supply chain
SCM	supply chain management
SCOR	supply chain operation reference
SCPM	supply chain performance measurement
SCPMS	supply chain performance measurement system
SEM	structural equation model
SMART	simple multi-attribute rating technique
SRM	supplier relationship management
TCO	total cost of ownership
TOC	theory of constraints
TQM	total quality management
VMI	vendor-managed inventory
WTM	work transformation matrix

PREFACE

The basic motivation for writing this book was to summarize the studies that the authors have performed while analyzing the exhaustive literature on supply chain management and supply chain performance measurement in the context of textile enterprises.

The initialization of economic liberalization has given the required thrust to the textile industry around the world, particularly in India, resulting in its emergence as one of the largest globally. At the same time, academicians, researchers, and practitioners have realized that the importance of supply chain performance is universal. They found that it is a necessity for a balanced approach in measuring the performance aspects in two ways, that is, within and across the industry boundaries.

In this book, an attempt is made to deal with the various aspects of supply chain performance measurement issues in the textile supply chain industries.

The main purpose of this book is to understand, implement, and evaluate the supply chain performance of the textile industry, considering the contribution of its cyclic processes. The purpose of the book is also to facilitate the textile industry with a comprehensive set of performance subcriteria to evaluate their supply chain performance measurement and to elaborate the supply chain performance practices in identified textile industries.

As the study is in its nascent stage, particularly supply chain performance measurement aspects in the textile industry, this study is undertaken to evaluate the various aspects of the performance of the supply chain. Delphi method is employed with minor deviation followed by the identification of companies of diverse sizes, and finally the analytical hierarchy process is applied to find out the crucial performance measurement criteria and subcriteria. Hence, in this study, four textile industries have been selected (two large-scale and two medium size enterprises). Around 80 industry personnel were contacted from four industries. The respondents were the general manager, the factory manager, the manager, the asst. manager, and the executives/supervisors involved in various functions related to the textile supply chain. Finally, the multicriteria decision-making technique was employed, that is, the analytical hierarchy process for executing pair-wise comparison and establishing priority weights of criteria, subcriteria, and estimating the contribution of supply chain cyclic processes to the overall supply chain performance. In this book, a research study is presented that uses the salient features of the balanced scorecard (BSC) by using its four perspectives (financial, customer, internal business process, innovation, and learning) and associated 23 (6, 6, 6, 5) subcriteria under these perspectives. The overall supply chain performance is affected by the performance of its cyclic processes, hence we selected procurement, production and distribution as alternatives after discussion with subject matter experts while employing this Delphi method.

This book aims to facilitate the textile industry by analyzing the use of a comprehensive set of performance criteria (BSC perspectives) and measures (subcriteria) and evaluates the contribution of each of the supply chain cyclic processes toward the overall supply chain performance, thereby enabling managers of independent units or the partners of the supply chain to improve the performance of the supply chain as a whole.

— Dr. Pranav G. Charkha
Dr. Santosh B. Jaju
Dr. Prasenjit Chatterjee

CHAPTER 1

SUPPLY CHAIN MANAGEMENT AND SUPPLY CHAIN PERFORMANCE MEASUREMENT

ABSTRACT

The concept of supply chain management (SCM) has evolved over the past few decades and has become an essential aspect of business management and operations. SCM involves the coordination and integration of various standalone enterprises within a SC to create value and meet customer expectations. Understanding value addition from an SCM perspective is crucial as it enables organizations to optimize their operations, enhance productivity, and achieve long-term competitiveness. By extending the enterprise beyond its boundaries and integrating it into the SC, organizations can gain a competitive advantage in a globalized market. The focus has shifted from individual organization competition to competition between SCs. To excel in this environment, firms need to effectively manage their SCs by emphasizing aspects such as SC integration, coordination, and supplier-buyer relationships. SCM is applicable in various industries, and it encompasses strategic processes such as supply management, manufacturing, and distribution. The successful implementation of SCM relies on building cooperative relationships among SC partnering firms and understanding the interdependency between organizations. The evolution of SCM has led to a shift in emphasis from internal improvements to external coordination and integration. The complexity of SCM requires a systematic understanding of its philosophy and continuous improvement efforts. Performance measurement is a vital aspect of SCM,

Supply Chain Performance Measurement in Textile Enterprises. Pranav C. Charkha, PhD, Santosh B. Jaju, PhD & Prasenjit Chatterjee, PhD (Authors.)

helping organizations assesses effectiveness and efficiency in meeting customer expectations. Overall, SCM plays a crucial role in enhancing an organization's productivity, profitability, and competitiveness in the dynamic business landscape.

1.1 SUPPLY CHAIN MANAGEMENT (SCM): INTRODUCTION

Supply Chain (SC) is primarily a group of standalone enterprises linked by the virtue of goods and services for which they create value autonomously or mutually to fulfill the expectation of the customer. SC works as an extended enterprise for an organization. The question is, what is the importance of understanding value addition from SC perspective? Why is it important to look at standalone enterprises with its SC extension? What are the aspects of managing SC for firms' success? What are important areas, where an organization must focus to excel in getting desired output? These are a few of the queries, which is important to deal with before heading toward questions dealing with "how to."

The last three decades witnessed immense development in the field of business management and operations management, etc. A lot of innovative methodologies and concepts has been evolved surpassing the conventional ones. The fewer new concepts are logistic strategic management, lean manufacturing, and management, agile manufacturing, business process reengineering, and balanced scorecard (BSC). Among these, supply chain management (SCM) is certainly a very result-oriented approach developed and applied in the organization around the globe.

The evolution of SCM is experienced and found in literature, which happens to be in early 1980. It was first coined by Oliver and Webber (1982) in an article published in *Financial Times*. SCM was first considered to be the solution for purchasing and cost-saving related concepts. SCM concept as we know and study came to extensive application in 1990 with issues such as SC integration, SC coordination, supplier–buyer relationships, etc. Later major development took place, and a significant increase in evolving new domain of SCM is witnessed. Subsequently, publication increased giving SCM the shape we observe today.

Thus, SCM is not very old, having vast theoretical background; however, it is rather new and at a very nascent stage of development. This idea has only been in use for roughly a decade and a half. As a result, there is a convincing response to the question "Why Now?" that may be traced to the impact of globalization on industries. By widening the environment and growing contest, globalization encourages consumers to raise larger requirements in front of service/product producers to be more quality conscious and increase their efficiency, effectiveness, and flexibility in managing optimal costs (Dangayach, 2005). Hence, enterprises are striving to achieve cutthroat cost, best quality, innovative technology deployment, and other competitive advantages by not only moving as standalone firm, rather integrating itself into SC and then competing with other SC. It thus means individual organizations have become now extent enterprises and more often SCs. Currently, one popular way to gain competitive advantage is to bridge the gap between independent units through SC efforts and competently performing numerous cyclic processes. For this, numerous producers are concentrating on its SCM initiatives (Deshmukh and Mohanty, 2011).

Hence, as we say, the scenario is shifting from "organization versus organization" competition to "SC vs SC." Sustainability in this volatile market mainly depends on the organization's competence in making a recognizable contribution to cooperate among SC partnering firms. The focus must remain on "how firms work during their interdependency."

In today's business management strategies and studies, managing the SC is a critical multifaceted topic. It is beneficial in increasing a company's productivity and profitability by using an innovative concept in business operations with long-term competitiveness (Gunasekaran, 2004). SCM has grown in importance in businesses that are supplying products customers. The consideration of stakeholders is of huge significance to firms; in spite of the volume/services firms are offering. Though, SCM is possibly most simply implemented in manufacturing. Also, SCM strategies are vitally important.

Manufacturing companies all across the world have embraced the SCM philosophy as critical and, at times, fundamental to their success (Dangayach, 2003). In the recent past, different manufacturing/service sectors in India have stepped to build up and perk up their SC strategic design, resulting in obvious performance enhancement (Swaminathan, 2006). They started incorporating new industrial engineering knowledge and philosophy to manage local and global competitiveness. However, there is a perspective difference among large-scale industries and small- and medium-scale industries in the implementation of these concepts and philosophies. Hence the difficulty is building up SC strategies, and the rest of the industries, especially medium and small-scale industries require complete wholesome rework in their management philosophy.

A classic SC involves an arrangement of raw materials providers, and producers of semi-finished, assembled, and finished goods collectively with a customer focus (Mentzer et al., 2001). An efficient SC delivers the right number of products/services at the right place on the right instance while optimizing the finances involved (Saad, 2002). In short, a typical SC is comprised of strategic processes such as supply management/ procurement, manufacturing/production, and distribution/transportation (Deshmukh and Mohanty, 2011).

SCM is also invasive everywhere. Not a single aspect of any business, where one cannot find the application of SCM. Let us consider the principle of quality management—an integral part of any organization management—and we must introspect whether it is possible to rate the quality of goods/services as measured by consumers, excluding the effectiveness and efficiency of suppliers and subsequent buyers. The answer will definitely be not at all. Until and unless all coordinating firms in SC offer significant cooperative contributions, value in business cannot be created.

1.2 DEFINING SC

Given the abundance of literature, one may conclude that many people have attempted to define SCM. Nonetheless, there is widespread theoretical agreement on the concepts of SC and SCM beyond any reasonable dispute. Table 1.1 summarizes some of the attempts made by previous authors.

TABLE 1.1 Definitions of SCM.

Sr. No.	Researcher/practitioner	Definition
1	Stevens, 1989	SCM is a system that involves material suppliers, production facilities, distribution centers, and customers linked together through the feed-forward flow of materials and reverse flow of information.

TABLE 1.1 *(Continued)*

Sr. No.	Researcher/practitioner	Definition
2	Berry et al., 1994	SCM attempts to build consumer trust through knowledge sharing, innovating great product ideas, and narrowing the supplier base to a single original equipment manufacturer (OEM) to free up management resources to focus on developing meaningful, long-term relationships.
3	APICS Dictionary, 1995	SC consists of two parts; (1) The process of connecting suppliers and users from the origin to the end consumption of the finished product. (2) The internal and external operations that allow the business model to manufacture and offer benefits to its consumers.
4	Global SC Forum, 1997	From consumers to original suppliers, who offer products, services, and information to the customer, the SC involves the coordination and integration of critical business operations.
5	Mohanty and Deshmukh, 2004	The smooth and cost-effective movement and storage of commodities, in-process stock, completed items, and relevant data are characterized as SCM.
6	Sople, 2012	SCM is an important business process that links the supplier to the customer through manufacturing. It is a network of facilities and distribution points. The network performance is the function of material procurement, conversion of these materials into finished goods, and finally distribution of finished goods.

Depending on facts, the authors would like to define SC and SCM as follows:

Supply Chain is nothing but a set of coordinating, cooperating enterprises/firms/business organizations creating and adding value through a series of operations/activities resulting in the generation of goods/services satisfying the expectation of customer and enhancing the market share of an enterprise.

Supply Chain Management is a set of strategic initiatives implemented to operationalize the effectiveness and efficiency in delivering the right product at the right quality at the right time to the right customer by optimizing the resources right from the procurement of raw material, conversion into usable goods/services and delivering to customer.

1.3 TYPES OF SCs

The concept of SCM uses complex terminology, thus restricting management from being a concept well understood and also for its effective practical implication (Ross, 1998). Understanding of "SCs" has been more familiar among authors than the concept of "SCM" (Lambert et al., 2005). La Londe and Masters (1994) defined SC as a set of enterprises that passes materials forward. Generally, various individual firms are involved in producing a product and delivering it to the end user in an SC—raw material suppliers, manufacturers, product assemblers, distributors, wholesalers/retailer merchants, and logistics providers are all SC partners (Lai et al., 2002). Another study describes SC as a network of functions that are involved in the creation of value in the form of goods and services delivered to end customers in the form of various processes and activities (Christopher, 2000).

SC can be progressively categorized into three types explained as follows:

(1) A typical SC
It is made up of three or more companies that are linked by intermediary functional processes like procurement, production, and distribution. Materials are procured, goods are manufactured, and services, finances, and information are disseminated from a source to a client, as explained through Figure 1.1.

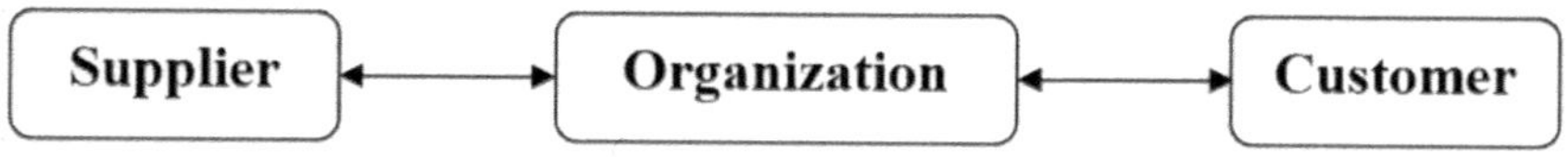

FIGURE 1.1 Typical SC.

(2) An extensive SC
It includes the direct supplier's supplier and the direct customer's customer, all of whom are linked by one or more product flows, product flows, operations, financial activities, and communication processes, as shown in Figure 1.2.

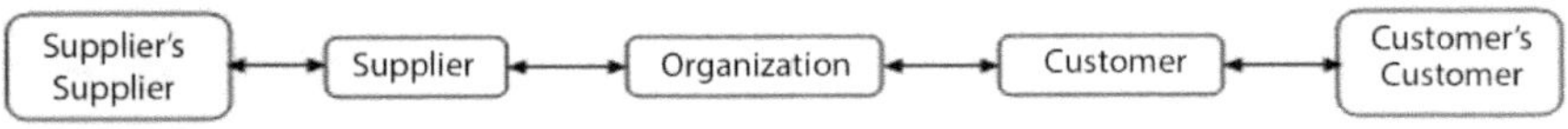

FIGURE 1.2 Extensive SC.

(3) A superlative SC

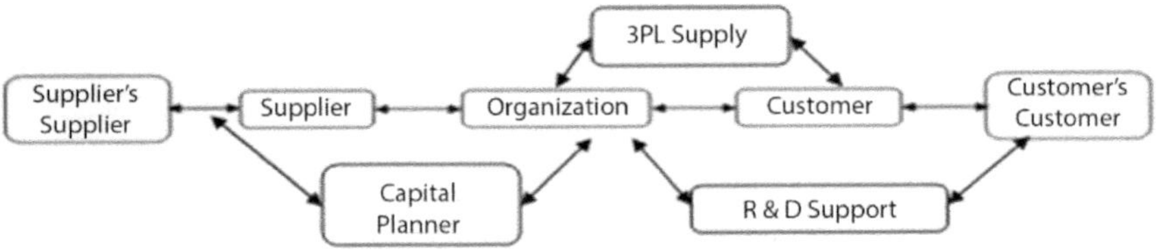

FIGURE 1.3 Superlative SC.

As shown in Figure 1.3, it encompasses all organizations involved in product flows, operations, financial activities, and communication processes from the first supplier to the last customer.

1.4 EVOLUTION OF SCM

The early 1980s was the origination period for SCM for manufacturing industries. Paradigm shift and innovation in three-letter philosophies such as just-in-time (JIT) (Vrijhoef and Koskela, 2000) and total quality management (TQM) (Wong and Fung, 1999) has developed the foundation of SCM. Just like various internal quality improvement programs aiming at enhancing overall competitiveness, SCM becomes an example of progression and collective improvement (Saad et al., 2002). The purpose is not only limited to boosting the efficiency of organizations but also has now been widened to include techniques of waste reduction and creating value along all the processes of SC (Harland et al., 1999). The application of SCM has shifted from the inner core to external cores or bridging/interaction processes between two organizations, based on coordination between the enterprise and its third parties employing close loop and cooperative learning. SCM is a set of procedures aimed at managing and coordinating the entire SC, starting with suppliers of raw materials to satisfying the consumers (Slack et al., 2001), resulting in improved synergy via the association of partners throughout the entire SC (New and Ramsay, 1997).

SCM, as defined by the Council of SCM, is the proper management of processes or functions involved in sourcing and procurement/supply management, conversion (production/manufacturing management), and distribution/demand management. It simplifies coordination and collaboration with channel partners, which might include suppliers, manufacturers, vendors, and customers. SCM unifies supply, manufacturing, and capacity utilization within and throughout businesses. As illustrated in Figure 1.4, the evolution of SCM can be discussed depending on the age of its development.

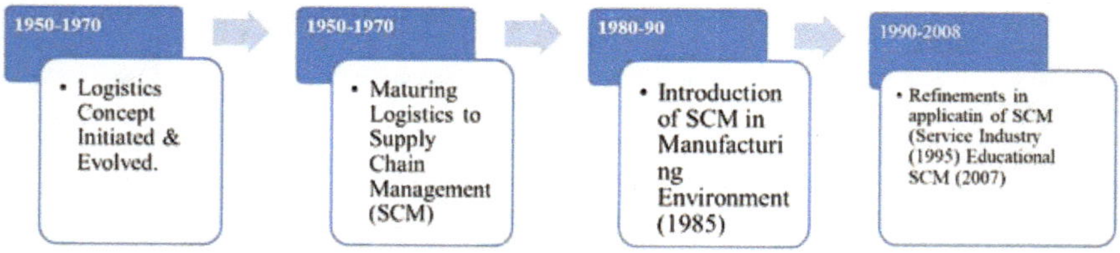

FIGURE 1.4 Evolution of SCM.

Chopra and Meindl (2007) stated that the functions of SC need to satisfy the customers' requirements ultimately. Along with suppliers and manufacturers, SC also consists of distributors, storage places, and customers. The functions generally undertaken by SC are forecasting, marketing, product design, process development, production, operations, financial management, and customer service.

Effective management of linking processes between organizations such as supply management/procurement (supplier–manufacturer), manufacturing (in the plant), and distribution/demand management (manufacturer–customer) is a major concern for SCM (Von Hipple, 1986), also likely important is the integration of both upstream and downstream processes (Christopher and Juttner, 2000). Result of such coordination and integration leads to more effective and longer-term relationships among suppliers, manufacturers, and buyers (Kosela, 1999).

Still, SCM is a long, multifarious, and vibrant process. Its successful accomplishment needs to be coupled with a systematic understanding of the philosophy itself (Whipple and Frankel, 2000). Its execution is also seen as being strongly reliant upon the capacity to create, control, and redesign associations among individuals, enterprises, and also networks along the SC (Spekman et al., 1998). This needs not only novel preparations and culture but also needs adequate dedication, facilities, and time to design.

However, is it mandatory to know that SCM is complicated and has proved to be intricate to apply? It is a multiactivity process, depending on close and long-term associations within as well as among firms (Saad et al., 2002). Challenges for the development of culture focused on peer learning, simplicity, and bond of conviction lead to the success of SCM. Sustaining and constantly enhancing the coordination and integration of all interfacing activities are the primary obstacles in order to improve the SC's overall performance, given the increased reliance on suppliers, the expansion in third-party contracts, and severe competition. Hence, strategies for continual improvement by performance measurement systems must be the base for efficient SCM.

Controlling an SC within the country is difficult due to various types of uncertainties in supply, process, and demand. In most developed countries, there are limited uncertainties in the accessibility of basic requirements for any kind of business such as power, roads, water, etc. However, in developing countries, infrastructure is not sufficient which poses several newer types of challenges. Especially, in the case of the process industry, SC typically begins supply management/procurement after which manufacturing/production and finally demand management/distribution of end products.

SCM has become a hot area of research and debate in recent years, after its introduction by consultants in the 1980s and later by academicians in the 1990s (Sople, 2012). SCM gains vast focus such that, it is not possible now to consider the aspect of manufacturing, marketing, distribution, customer management, or transportation independently without notice of the advancement of SCM or concerned topics (Ross, 1998).

1.5 EVOLUTION OF SC PERFORMANCE MEASUREMENT

Performance measurement as coined by Kennerley et al., 2002 is the process of quantification of effectiveness and efficiency of action. Meeting the customer's expectation is a measure of effectiveness, whereas, efficiency deals with how a firm provides products/services to customers to meet their satisfaction, that is, at what economic level, with what degree of flexibility, and with how much time spent on them. Hence, performance measurement systems are explained as comprehensive sets of measures used to quantify the performance level of action.

Sink and Tuttle (1998) asserted that "what you measure, you can manage," and suggested a designing performance measurement system for process improvement. The proposed step for designing a performance measurement system as mentioned below:

- Preparation
- Determination of what to measure
- Determination of measurement techniques to process data
- Choosing what and how to collect
- Process and output validation
- Linkage of measurement to improvement
- The feedback system for improvement initiatives

The ultimate objective of the study is to achieve a competitive advantage by improving the overall performance of textile SC by evaluating an SC (Angerhofer and Angelides, 2006).

There have been few attempts to construct models for evaluating the performance of SCs (Gunasekaran and Kobu, 2007). A lot of critics exist over the most suitable way to categorize them. From the review of various work done, the present SC performance measurement frameworks can be classified by the considering following three main aspects that are mentioned commonly by other scholars: Strategic perspective or outcome-based aspect (balanced scorecard) (Kaplan and Norton, 1992, 1996, 1997), process aspect (SC operations reference, SCOR model), and frameworks (SC Council, 1996).

Most of the studies focused often on a single area of performance measurement, preferably cost, whereas research on a comprehensive inclusion of all perspectives for the performance evaluation of SC from all the stakeholder aspect has been proportionately neglected (Beamon,

1999; Gunasekaran et al., 2001; Chan and Qi, 2003; Otto and Kotza, 2003; Huang et al., 2005; Aramyan et al., 2007; Charkha et al., 2014a, 2014b).

Some year-wise development in the studies of SC performance measurement is depicted in the Figure 1.5.

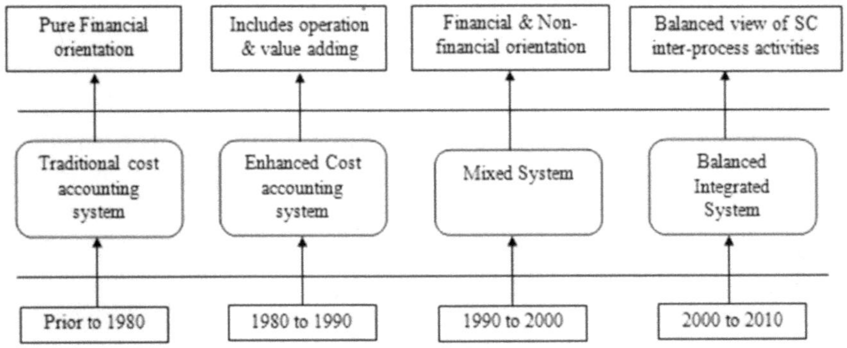

FIGURE 1.5 Periodic evolution of SCPM studies.

1.6 SCM IN INDIA

It is well understood that SCM has a lot of challenges even in developed nations such as the USA. Developing nations such as India are experiencing even more complex challenges. Multinational organizations can deploy strategic SCM initiatives, which are probably difficult to adopt in India. One of the most important hurdles in managing SCs in India is probably the intellectual infrastructure, the infrastructure required for shipping, power sources, and water sources. Another aspect is government taxation policies with reasonable warehouse management, distribution channel facilities make efforts arduous to effectively and efficiently coordinate and manage SCs. Conversely, only major industrial players are well conversant with the application of information technology and knowledge management techniques; however, their relatively smaller coordinating partners do not have much orientation about the latest trends and techniques of management information systems. This makes it difficult to build more responsive SCs. Following constructive efforts, such technologies, and concepts can easily be implemented in organizations and which no longer a remote task. Most organizations have started implementing SCM.

In recent years, India has seen increased use of the Internet for processing customer orders, customer support, production planning, and other purposes, making SC stronger and more responsive. Radio frequency identification (RFID) and global positioning system (GPS) are now frequently used concepts adopted by Indian logistic players, which gives a clear and transparent view decision making for in-process inventory in their SCs.

However, another aspect of concentration is the application of quality engineering and quality management practices in the industry to make SC more effective and efficient. Practices such as JIT, vendor-managed inventory (VMI), KANBAN, Kaizen, etc. Concepts such as JIT have been well understood by bigger industrial players, whereas, looking at uncertainties in operating and managing SCs in India, firms are evolving realistic inventory management systems to reduce inventory and associated investment and expenses. The government has practically taken a lot of steps in building a robust economy for the smooth operation of SCs.

SCM that is effective and efficient allows the company to grow in a turbulent environment. Section fetches the conceptual idea of SCM at the same time importance of associated terminologies for ensuring the improved performance of SCs in India. To ensure performance to expectations, business firms should address the application of techniques/concepts that strategically facilitate an increase in productivity, profitability, and market share. SCM has proven its benefits in strategically facilitating improvements.

Since the 1980s, the interest of researchers, practitioners, academicians, and industries has increased in the application of SCM looking to observe its benefits in cooperative and collective business associations. SCM has periodically gained importance in India (Vrat, 1998). Indian business organizations need to look beyond their conventional way of dealing with their counterparts due to various competing parameters such as; ever-increasing variety in products, globalization, short product life cycles, and uncertainly in markets.

Indian firms are taking substantial initiatives and are not at the cutting edge to implement SCM practices. Best-in-class SCM methods are being observed in successful businesses. India has tremendous potential to be the world-class source of products and a world-class market as well, which is relatively less explored. In the form of MNCs, we can see the tip of the iceberg in the last decade for SCM recognition. The potential capacity of the economy driven by micro-, small-, and medium-scale enterprises together with SCM practices must also be realized. At the same time, factors such as government policies, growth rate, development of rural India, and last but not the least unstoppable growth and change in mindset of Indian firms.

1.7 PERFORMANCE MEASUREMENT SYSTEM

1.7.1 DEFINITION

Nowadays, the performances of business firms are encountering a lot of issues due to volatile changes (Yusra et al., 2015). Decreasing performance might pose serious consequences to businesses, even sometimes shut down (Sorooshian et al., 2011).

Performance measurement (PM) is a concept that is frequently elaborated on but hardly ever defined. PM is, in its purest form, a process of quantifying activity, where measurement equals quantization and activity equals performance. Business organizations have achieved their strategic objectives by performing and fulfilling their consumer's expectations with enhanced efficiency and effectiveness compared to their counterparts. Efficiency and effectiveness are terms very specifically defined and applied in the PM system. Efficiency is measured in terms of how well resources are used to meet customer expectations, whereas efficiency measure is associated with how well resources are used to meet customer expectations. Performance measurement has been and will always be an important aspect as it not only recognizes two important parameters but also point out much-targeted courses of action.

To elaborate more, let us assume the case of one of the parameters of quality performance—the reliability of the product. In relation to this parameter, effectiveness relates to attaining the maximum degree of product reliability which in turn leads to the achievement of customer satisfaction. Simultaneously, efficiency relates to the optimization of resources, by reducing failure in the system, reworking, etc. For this reason, it is rightly said that effectiveness and efficiency have a big role to play in a firm's performance. Hence, it is well defined as follows:

- ✓ The practice of quantitatively measuring the efficiency and potency of action is known as performance measurement.

- ✓ A performance measuring system is a collection of indicators that are used to quantify the efficiency and effectiveness of actions.

In a nutshell, performance measurement covers a wide range of topics. We will go into greater detail about establishing performance measurement systems, specifically for SCM, in the next sections. The remaining chapters have been divided into four divisions, namely (1) Individual performance measurements, (2) performance measurement as an entity, and (3) the link between the performance measurement system and the environment in which it functions can all be analyzed at three levels. Take, for example, the "set of measurements" in Table 1.1, which are employed by a major US supplier of computer-based information processing equipment. This performance measurement system can be examined at the individual level by asking the following questions:

- What performance measures are used?
- What are they used for?
- How much do they cost?
- What benefit do they provide?

Performance measurement as coined by Kennerley et al., 2002 is the process of quantification of effectiveness and efficiency of action. Meeting the customer's expectation is a measure of effectiveness, whereas, efficiency deals with how a firm provides products/services to customers to meet their satisfaction, that is, at what economic level, with what degree of flexibility, and with how much time spent on them. Hence, performance measurement systems are explained as comprehensive sets of measures used to quantify the performance level of action.

Sink and Tuttle (1998) asserted that "what you measure, you can manage," and suggested a designing of performance measurement system for process improvement. The proposed step for designing a performance measurement system as mentioned below:

- ➢ Preparation
- ➢ Determination of what to measure
- ➢ Determination of measurement techniques to process data
- ➢ Determination of what to collect and how to collect
- ➢ Process and Output validation
- ➢ Linkage of measurement to improvement
- ➢ The feedback system for improvement initiatives

1.8 SC PERFORMANCE MEASUREMENT

1.8.1 DEFINITION

SC performance measurement can be defined as a method of evaluating the system's performance. SC performance measurement can be accomplished at three distinct levels of decision making which can be broadly categorized as;

- ✓ Strategic: High-level decision-making regarding the marketplace to be directed to the size as well as the location of production sites, the partnerships to ascertain with suppliers, etc.
- ✓ Tactical: Focus on adopting measures (best practices) that result in efficiency in SCM such as demand planning, purchasing strategy, etc.
- ✓ Operational: Regulating the real flows through different ends of the SC (taking orders from customers, transmitting the information downstream, controlling the costs, ensuring logistics operations, etc.).

SC performance measurement can be successful, only when it encompasses the comprehensive strategic perspectives of an organization such as financial aspects, customer orientations, internal operations, and lastly growth opportunities.

Holistic SCPM must consist of performance measures must be a mix of both qualitative and quantitative nature.

1.8.2 NEED OF SCPM

Studies pertaining to the development of performance measurement frameworks for SCM were dominantly conducted in developed countries. Accordingly, few efforts have been made in developing countries to study the SCPM concept specifically in India. Whereas, the application of SCM practices is globally accepted. In addition, SCM has emerged as a vital concept in various developing countries. Consequently, the present work focuses on an SCPM framework with special reference to the textile industry in India.

Research in this thesis proposed to design a generic strategic SC performance measurement framework and validated using cases of four textile industries. This framework developed uses the perspectives—financial, customer, internal business operation, innovation, and learning—of a balanced scorecard including SC cyclic operation, integrating upstream suppliers (sourcing/procurement), and downstream partners (distribution/delivery) with value creation units (manufacturing/production). The importance of the BSC perspective is that it gives an overall view of an industry covering the perspective of all stakeholders, such as the financial perspective that gives a management view, the customer perspective that gives end-user expectation, an internal business perspective that shows efficient functioning of plant and innovation and learning stressed on imbibing best practices. As a result, the projected framework can evaluate the performance of textile SC in a very holistic nature. The proposed framework facilitates business firms to make proper SCM decisions considering these four aspects of the industry and by evaluating these perspectives and associated attributes on three SC cyclic processes, namely, supply management/procurement, manufacturing/production, and distribution/demand management. Measurement of SC performance is done by using specified subcriteria under each perspective.

1.8.3 SCOPE OF SCPM

SC performance measurement (SCPM) is very wide and cross-boundary spanning. SCPM plays a vital role in measuring and managing the efficiency and effectiveness of any organization's SC. Decision making in any business firm's SC is a matter of inclusion of a variety of parameters, which is a relatively very critical and complex process. SCPM solicits intense benefits for managers and facilitates their decision making. Customer's expectations, firm's aspirations, and general performance can be well understood using comprehensive SCPM. SCPM also reveals the effective and commanding controller of the organization's SC. SCPM enables organizations to work as extended enterprises. It helps in strategically framing a firm's objectives and goals (both short- and long-term objectives and goals) and procedure of accomplishment. Additionally, the performance measurement process envelopes all SC operations of an organization. SCPM ensures integration and coordination among SC partners.

KEYWORDS

- **supply chain management**
- **value addition**
- **competitive advantage**
- **supply chain integration**
- **supplier–buyer relationships**
- **performance measurement**
- **continuous improvement efforts**

CHAPTER 2

SUPPLY CHAIN MANAGEMENT: REVIEW OF TEXTILE ENTERPRISES

ABSTRACT

Supply chain management (SCM) is of utmost importance in the textile industry, facilitating the efficient and effective movement of materials, information, and products throughout the entire SC. This abstract provides a summary of SCM in the textile industry, highlighting its global presence, its significance in India, and the practices employed in textile SCM. The textile industry has a global presence, with SCs spanning multiple countries and continents. This global nature presents unique challenges such as managing long-distance logistics, ensuring compliance with diverse regulatory frameworks, and coordinating with a wide range of stakeholders. Effective SCM practices are essential to overcome these challenges and optimize the textile SC on a global scale. India is a prominent player in the textile industry, with a rich heritage and a robust manufacturing base. The textile industry in India is characterized by a complex SC involving various stages, from raw material sourcing to fabric production and garment manufacturing. SCM plays a crucial role in managing the diverse and geographically dispersed textile SC in India, ensuring timely delivery, cost efficiency, and quality control. Textile SCM practices encompass a range of activities such as demand forecasting, procurement, inventory management, production planning, logistics, and distribution. Efficient SCM practices in the textile industry involve collaboration and coordination among suppliers, manufacturers,

Supply Chain Performance Measurement in Textile Enterprises. Pranav C. Charkha, PhD, Santosh B. Jaju, PhD & Prasenjit Chatterjee, PhD (Authors.)

distributors, and retailers, leveraging technology for real-time visibility and data-driven decision-making.

2.1 SCM IN THE TEXTILE INDUSTRY

Last couple of decades, supply chain management (SCM) made coordination, and collaboration among buyer and supplier organizations firms have been exhaustively explored to ensure effective and efficient performance of the entire chain. It has also gained extensive interest from both industry experts, academicians, and practitioner communities.

Supply chain management, as defined earlier, is a link of value-adding activities to ensure the satisfaction of the customer. It epitomizes a vital shift in the practices of enterprise management. SCM is the most efficient ways to strategies enterprise performance management. It conveniently incorporates suppliers, manufacturers, distributors, and customers. The ultimate objective is to improve the firm's performance as well as the supply chain as a whole in a current volatile market.

Any supply chain (SC) has two parts, namely, the upstream part and downstream part. Their effective coordination and integration lie in successful SCM implementation. This will lead to higher satisfaction of the customers and improved organizational performance. Studies have proven that SCM is a vital player in improving an organization's performance and a strong driver for achieving competitive advantage. To make this happen, the responsibility of supply chain managers has been very important in integrating the reverse flow of the activities (from the customer end to procurement of raw materials).

Many SCM studies were conducted in developed and developing countries. Existing work by various authors was studied for compiling this book. Studies referred to were categorized based on the type of work done in a particular paper.

Identified papers were grouped into four categories as per the work conducted in them:

i. Review papers
ii. Research papers
iii. Case study papers
iv. Others (conceptual, etc.)

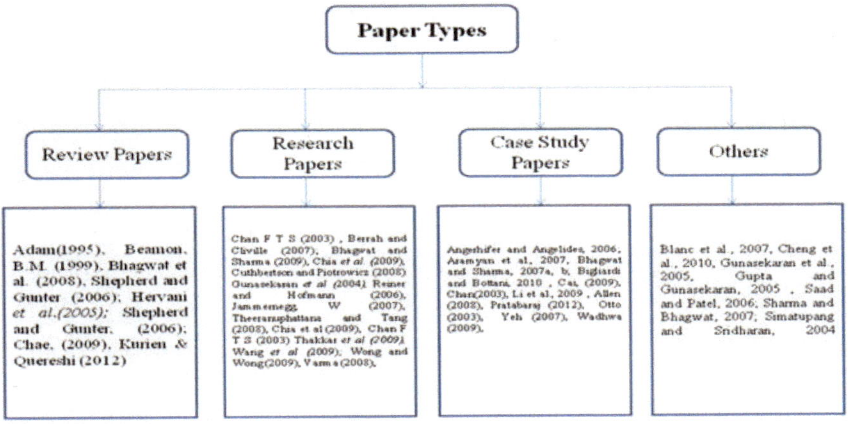

FIGURE 2.1 Classification of identified papers.

During the literature review, it was found that most of the studies were conducted in developed countries, hence we decided to see the list of countries more concerned about such studies. Hence, a collection of studies conducted in various countries and industries focused are as shown in Figures 2.4 and 2.5, respectively.

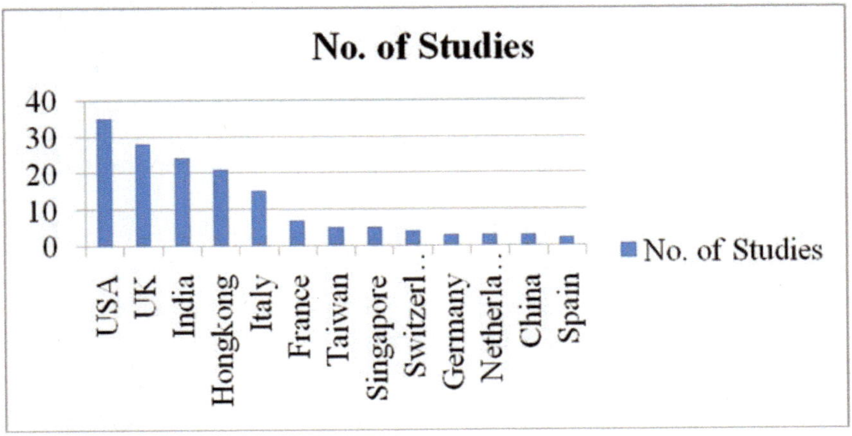

FIGURE 2.2 Country-wise number of studies.

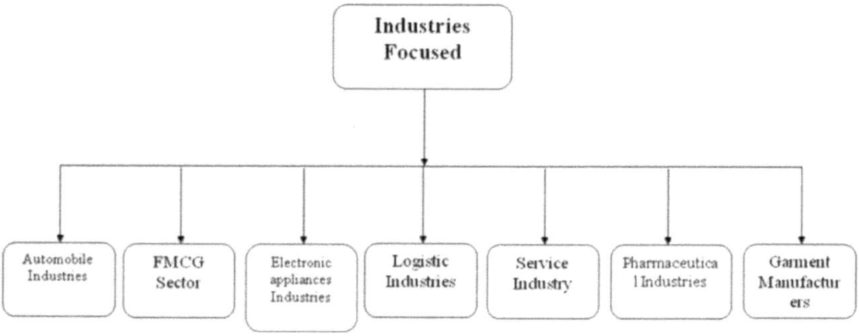

FIGURE 2.3 Industries considered for supply chain performance measurement.

2.2 TEXTILE INDUSTRY—GLOBAL PRESENCE

The textile industry is a major contributor to the world's economy. A variety of materials are being produced and sold by industry such as cotton, yarn, fiber, and finished products or apparels. It is believed that the global textile industry contributes 5% of the compound annual growth rate (CAGR) till 2020. It is also speculated that the textile industry market may cross the US$800 mark. Similarly, it is predicted that the apparel and allied product market will cross US$1000 billion in the coming few years. Owing to embracing the domestic textile product, the textile industry with its global presence may get affected and rise and fall.

The textile industry is one of the fastest-growing sectors with key players competing globally are nations such as China, European Union, the United States, and India. Brief country-specific statistics are as shown in Figure 2.4.

Textile industry is a fastest growing sector. Sector is having strong competition among countries China, European Union, the United States, and India. China one of leading player has risen their exports by approximately 3% by year 2018, sharing almost one-fourth market among textile industry across the globe having export value of US$100 billion. Another leader in the textile industry is the European Union (Germany, Spain, France, Italy, and Portugal) sharing a value of roughly one-fifth of the textile industry worldwide while having a value of US$160 billion. Third in the league is India holding US$30 billion as export value; also contributing 6% production of total textile production globally. Regarding the dominance of these countries, the USA is the prominent consumer of textiles.

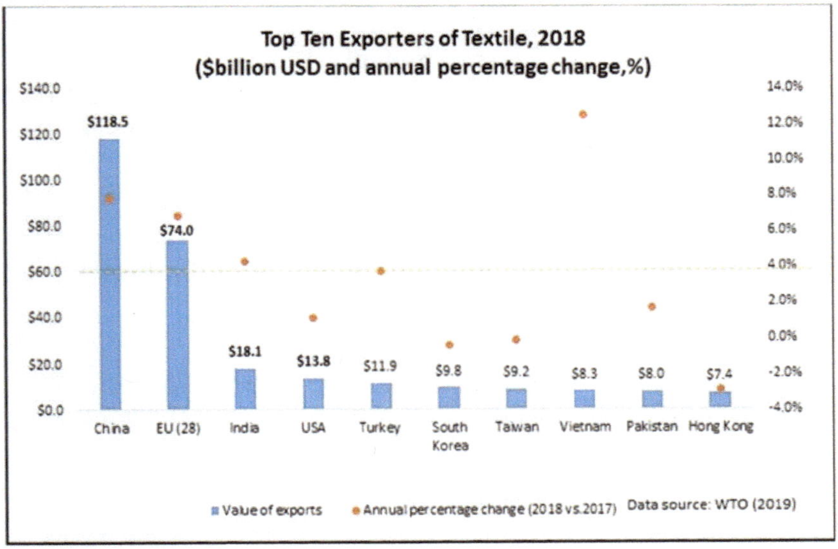

FIGURE 2.4 Top 10 exporters of textiles (2018).

(Source: https://medium.com/@jashimohan01/global-textile-industry-recent-trends-in-the-market-45d2d2b86392)

Having seen the downfall, the textile industry in Asia has been rapidly evolving rapidly and capturing the global market. In the past few years, Asia's textile industries have taken huge shape and extend its own boundaries using effective supply chain management.

Customer taste and clothing patterns has given much needed boost to Asian textile industry postrecession. Industry has adopted various techniques, technologies, and concepts to become modernized and sophisticated to serve customers to the full extent.

Even the growth of technical textiles has paced up very rapidly. The technical textile consists of mattress covers, bags, tents, parachutes, etc. Customer interest in purchasing innovative products has triggered a boom in this industry. In the near future, it is foreseen that this sector can grow by approximately 20%.

However, there are lot many challenges; the sector is facing which include cost, quality, variation in the market, etc. As enterprises enhance their avenues beyond the local markets enabling them to efficiently manage the outsourcing, as a result of which, respective supply chains are crossing the boundaries of countries. For example, any product is being consumed worldwide, while they are designed and manufactured using global inputs. Activities and part of logistics are outsourced on account of cost optimization. The current scenario of the textile industry of different countries of Asia is shown in Figure/Table.

TABLE 2.1 Features of Major Asian Textile Player.

Sr. No.	Country	Textile industry's features
1	India	✓ Having Agrarian economy. ✓ Textile trade is oldest and well settled. ✓ Contributing to nation's economy. ✓ As per reports, Indian textile industry was valued at US$150 billion in 2016–2017, and was estimated to reach US$250 billion by 2022. ✓ Varied sectors consisted of hosiery, power looms, and knitting plants. ✓ 62% of production is contributed by the power loom sector of the textile industries. ✓ Other subdivision consist of woolen textiles, silk textiles, jute, handicrafts, yarn, handloom, etc.
2	China	✓ Shaoxing is major textile cluster in China. ✓ Cultural center and also origin of silk making companies. ✓ As of 2014, and as per WTO, the exports of Chinese textiles and clothing reduced from US$236 billion in 2014 to US$206 billion in 2016. ✓ In 2015, the export value of textiles was US$11.53 billion. ✓ Country has around 100 textile companies having yearly sales of more than US$30 million.
3	Bangladesh	✓ Cotton and jute processing are the most prominent. ✓ Crucial economic sector in Bangladesh and a vital player in building the economy. ✓ Generates 80% of the country's exports and foreign exchange earnings. ✓ Textile sector houses an operational plant from spinning and weaving to knitting and dyeing. ✓ Industry encompassed the production and sale of finished garments and technical textiles.

TABLE 2.1 *(Continued)*

Sr. No.	Country	Textile industry's features
4	Turkey	✓ In recent past, Turkey's industry exports were valued at US$21 billion. ✓ Leader in the manufacturer of denim in the world. ✓ Fourth largest exporter of home textiles. ✓ Till last decade, the export estimated to 3.7% of global exports. ✓ Till the previous decade, Turkey was the fourth contributor to exports of apparel, while was seventh among textiles. ✓ To celebrate Turkey's 100th anniversary in 2023, it is targeted to emerge as a rapidly growing economy and be among top 10. ✓ Focusing on US$500 billion in exports. ✓ US$100 billion for the use of advanced technologies is being invested to achieve the target.
5	Thailand	✓ Thailand's has 4233 mills comprising of apparel mills, knitting mills, weaving mills, dyeing and printing mills, spinning mills, and manmade fiber mills. ✓ Major role player sectors for employment. Generates one million employees on an annual basis. ✓ Country's textile sector has an export value of US$7.55 billion during the last decade. ✓ Textile sector exported US$7.52 billion worth products.

Asia possesses the credits to be an emerging player in the textile and apparel industry. India, Bangladesh, and China are major contributors to textile manufacturing and trade. The Asian textile industry is prosperous and would be competing with bigger economies. Asian textile industry has extended their enterprises and bridged gaps with corresponding counterparts using effective and efficient supply chains. The coming decades will be the period of the Asian textile industry.

2.3 TEXTILE INDUSTRY IN INDIA

Indian economy in its past has very long and strong presence of textile industry. At present also, textiles sector is prime most contributors to India's exports sharing almost 11%. It is labor intensive industry as well major employment generator. The textile sector in India comprises of two segments as shown in Figure.

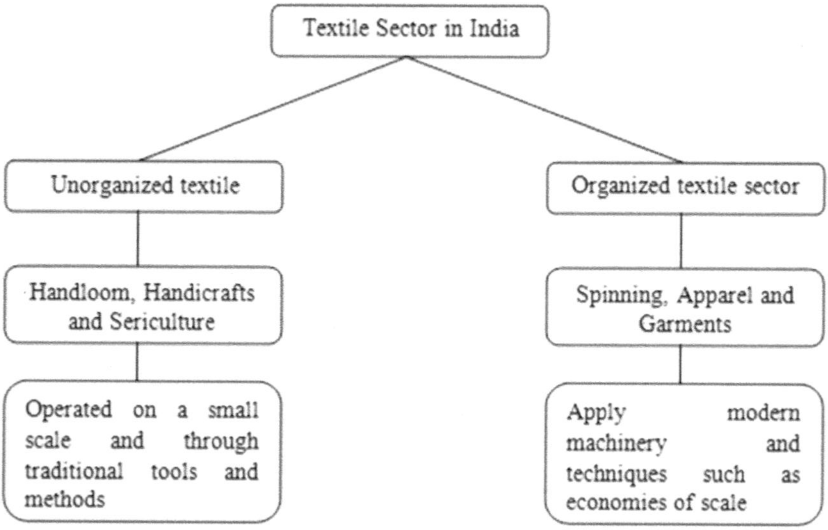

FIGURE 2.5 Textile industry segmentation.

Current statistics shows that Indian textiles sector, approximately valued as US$108 billion, forecasted to climb up to US$223 billion by 2021–2022. As mentioned earlier, the biggest employment generator next to the agriculture sector, recruits approximately 45 million personnel directly and 60 million indirectly. The sector is contributing 5% to GDP and 14% to the overall Index of Industrial Production (IIP).

The report claimed that the Indian textile sector possesses prospects to achieve US$500 billion in size (Wazir Advisors and PCI Xylenes & Polyester). It can be achieved with an increase in domestic sales from US$68 billion to US$315 billion, simultaneously enhancing export from US$41 billion to US$185 billion.

Indian textile sector is among the faster growing textile sectors globally, but not considerably focused on the advent of new technology and philosophies. Although the sector remained predominantly an unorganized industry for a large period of time in the past few years, but the scenario changed after the advent of economic liberalization of Indian economy in 1991. The ongoing philosophy of globalization helped the Indian textile industry to grow, which has now effectively become a prominent global leader. The textile supply chain in India is the most complex and fragmented supply chain starting with raw material production, procurement, production of finished fabric, and shipping to pre-determined locations through effective distribution channels. Figure 2.6 will show the components of the textile supply chain. Hence, it needs efficient and effective supply chain management. Most importance aspect that requires more focus to have global competitiveness of the entire SC is the integration of respective processes of textile industry (Chandra et al., 2000).

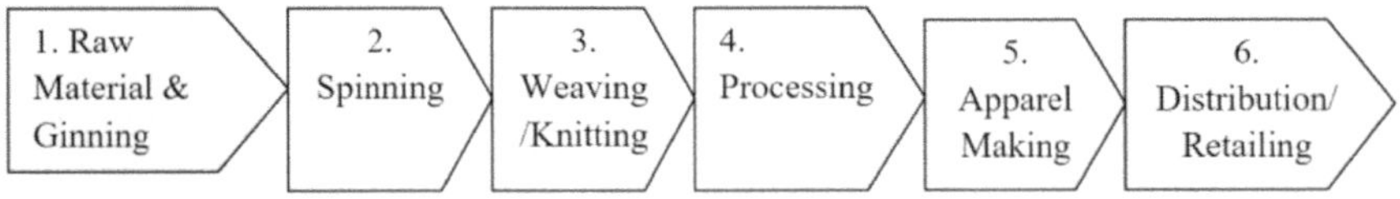

FIGURE 2.6 Textile supply chain.

The effect of trade agreements of the World Trade Organization (WTO) left remarkable boost in imports during last decade. Developing countries are producing textile products, consume them, and also provide them to developed countries for further use/processing situated miles away from the production base. To cope with the uncertainties of long-distance services and even for effectively monitoring the various processes, the concept of supply chain management has received extensive acceptance. Firms operating with different production philosophies and different products are collaborating to enhance competition, and evaluating the functioning of SCs has emerged as vitally responsible. Number of challenges are posing serious threats to the growth of the Indian textile industry in terms of government policies, fragmented structure of the chain with the control of the small-scale sector supplying raw material (ginning product/spinning product, less available foreign investments, logistical disadvantages, supply chain integration and effectiveness achievement and efficiency enhancement, unavailability of performance evaluation of supply chain of the industry and also less awareness for employing innovation and best practices to improve quality and competitive sustainability.

To succeed the SCs, firms must evaluate the performance of the various processes collectively rather considering the activity independently and identify areas of strategic improvement for increased supply chain efficiency and effectiveness.

Satisfying the above issue, the most important question for this research study was a list of "probable performance measures affecting textile SC and useful to achieve more competitiveness?" The first is that companies are achieving this through properly managing different processes of the supply chain. To estimate the value of the supply chain's processes, supply chain performance management (SCPM) systems are being utilized. During the textile industry's consideration, most of the upstream portion (supplier) of the supply chain is unable to decide on the appropriate measure. Hence efforts of these upstream companies to improve supply chain performance should be supported by their downstream SC partners—fabric manufacturers and customers. For example, the Indian textile sector is striving to improve financial competitiveness. On the other hand, customers are more inclined towards enhancing functioning of production pattern? Or asking for improvement in innovativeness, quality, or lead time? Hence, the overall perspective of an industry needs to be considered. It may either through innovation and learning aspect that tells you about the knowledge, skill, and system that your employees need, to innovate and build the right strategic capabilities and efficiencies for our internal business processes/operation, which delivers the desired value to the customer and eventually leading to high stakeholders' profit creation. This is achieved by using a balanced scorecard approach. Also, merely, focusing on the upstream partner could not lead to the comprehensive development of a performance measurement system, on the contrary, consideration of fabric manufacturer, and distribution system, along with supply management/ procurement, should be considered.

In view to achieving set targets to enhance the supply chain competitiveness of the Indian textile industry, an attempt is made to study the textile industry keeping SCM, and SCPM system's practices in view. The specific aim and research objective were set to design a comprehensive performance measurement framework for textile supply chain in India using relevant criteria, subcriteria, and contribution of supply chain cyclic processes towards supply chain performance depending on the issues of the industry.

2.4 TEXTILE SUPPLY CHAIN MANAGEMENT PRACTICES

The supply chain in textile industry is relatively complex in nature and perhaps long with number of enterprises involved (Jones, 2002). Below is the flow chart showing various participants in the textile supply chain. Accordingly, effective coordination of supply chain management within and among the partner's organizations (Forza and Venelli, 1997) is shown in Figure 2.4. This is required to reduce the lead time and enable the quick response approach need to adopt in changing market demands.

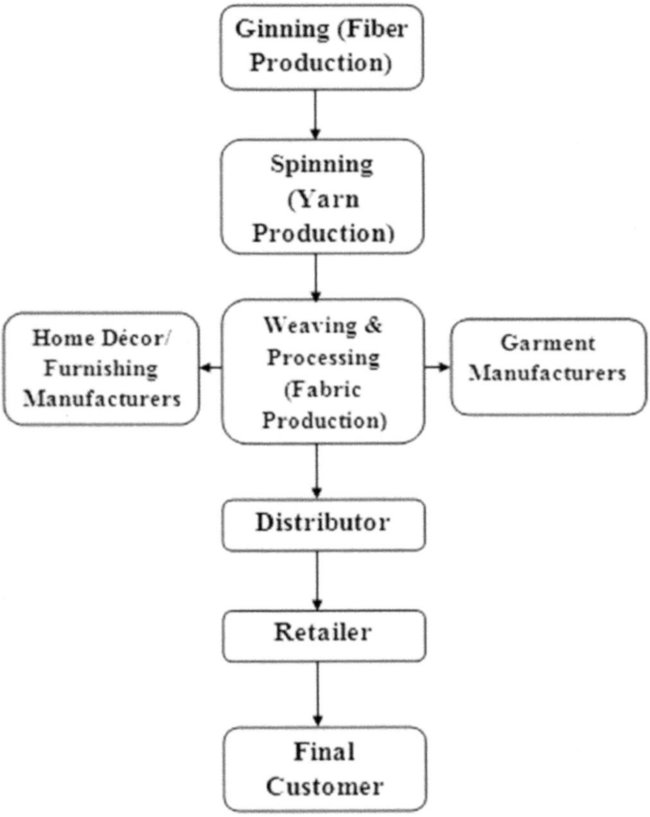

FIGURE 2.7 Participant in textile supply chain.

The quick response approach is the well acknowledged initiative of supply chain (Lumas and Vokurka, 1999). However, as textile supply chain cannot be compared with service industry supply chain such as medical supply chain or wealth supply chain management, etc. Effective balance in such company would be achieved by avoiding the bottlenecking phenomena, which makes it complicated to adopt quick response approach in such complex supply chain. Various configurations can be used to design effective and efficient supply chain. One of them is shown in Figure 2.5.

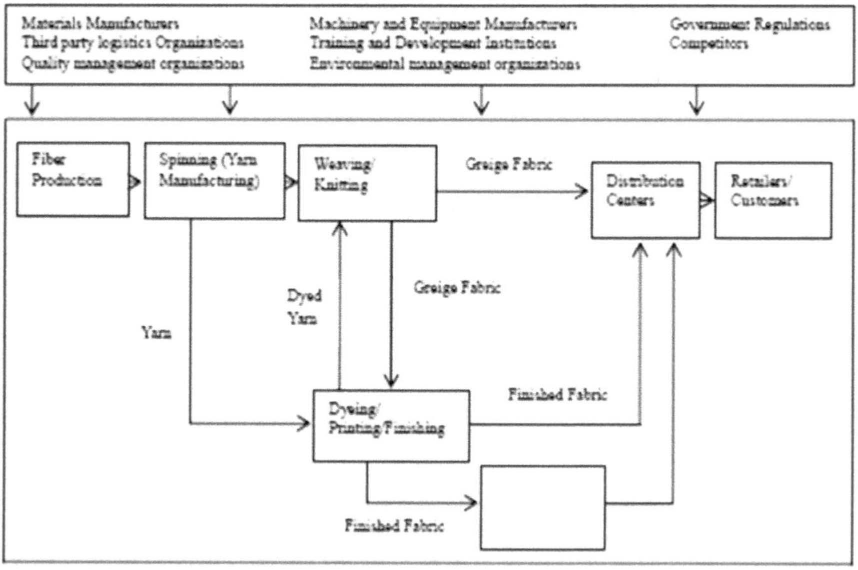

FIGURE 2.8 Textile supply chain configuration.

Textile industry is much dependent on its sourcing/procurement strategies which affect the performance capacities of the raw material supply base. There are a number of possible suppliers that may have differences in cost, lead times, and flexibility of their services. Suppliers having low-cost raw materials generally do not provide flexibility and cost-effective aspect, whereas, they need longer shipment times of around a number of weeks and frequently require that the entire production capacity be distributed consistently throughout the year. There may be suppliers with shorter lead times and allows flexibility, hence called responsive suppliers. Furthermore, other suppliers may be prepared to accumulate restricted amounts of raw material for a time of need for delivery (Agrawal et al., 2002). Hence, "procurement strategies are the combined model of synchronized decisions which needs to undertake for the formulation, improvement, and deployment of procurement resources to facilitate the attainment of the strategic objective of SCM framework" (Ukalkar, 2000).

As textile industry is configured by frequent changes in demands, industries are required to fulfill customer desires on time, with quality production and cost-effective delivery. Meeting the customer demands in terms of products and services, with the best quality, optimum price, and the minimum time is considered the key factor to sustaining competitiveness (Gunasekaran and Ngai, 2007; Leung, 2000). Frequent procurement and replenishment activities are main key factors for developing nation like India to be successful in global supply chain management (Gunasekaran and Ngai, 2007). Hence, the selections of appropriate suppliers are especially decisive in the case of textile industries.

Textile Industry, is an area where the cost and the quality of product are very important factors; conversely, existing competition is not dependent on standalone firm but rather consider the entire supply chain including activities from procurement to distribution of product or service. Achieving optimum cost and the best quality of the final product clearly needs focused efforts of inter-connected stages/functional divisions of the textile supply chain which are procurement, production, and distribution network. Further, quality may or may not be the constraint for raw material or finished products, but rather it takes into account the consideration of the operational process which adds value to final products. Hence, production operation scheduling and flexibilities are of great importance in textile industries as a number of processes through which the raw material is passing to get the final aesthetics is very large for, for example, spinning, combing, re-combing, weaving, processing, and finishing, etc.

Managing distribution decision are little difficult in domestic market which becomes more complex in international market due to their complicated and diversified nature. The issue of deciding the right distribution strategy is a major concern for almost major textile players in developing countries. This issue is important if companies need to market their product in an international maze of channels and need efficient domestic distribution planning which is yet to get strengthened.

Any distribution channel must be viewed as a sequence ahead of just marketing the product to its ultimate consumers. It should be an effort of a team working for a common strategic objective and goals (Czinkota and Ronkainen, 2007). Designing an optimal supply chain distribution strategy is very important as it affects the cost aspect of the product, serves the ultimate customer and needs good research studies to find the optimum distribution channels most importantly, it is resultant of the efficient internal business process.

In the recent past, there have been many studies conducted on the approach of quick response (QR). The importance of QR is much recognized by the textile and apparel industry in consideration with the organizational, managerial, and technological provision needed to achieve it. The importance of information technology across the bridging processes among supply chain partners is the result of this approach (Forza and Vinelli, 1997). Perry et al. (1999) explained the process as part of the QP approach implementation in the textile, clothing, and footwear industries. Still, the concept of QR is not well defined among each partner of the supply chain. On the contrary, it should be well conveyed and implemented to functional processes of supply chain such as procurement, production, and distribution.

More studies have been undertaken for garment manufacturers in the textile–apparel supply chain. Studies conducted on the US manufacturers have shown some significant parametric differences in the company's characteristics which include, product fashion category, fabric manufacturer delivery performance, relationship with suppliers, and retail customer with their relative size (Lee and Kincade, 2003). Comparative evaluation is done for Indian apparel manufacturers with the rest of the world (Bheda et al., 2003)

The model is suggested to promote sourcing centers as creating skills in SCM (Leung, 2000). Chandra and Kumar (2000) proposed a systematic system evaluation technique to control textile supply chain logistics. There exists a gap in the concept and practical implementation. Relationship management has been of crucial importance to envelope the management of chains and developing partnerships among various parties of the supply chain. Throughout the literature, it has been observed that collaborative relationships and partnerships are preferred philosophies for the benefit of all parties involved. The textile industry preferably dominated by a large number of retailers with discrete outlets, also, the manufacturing companies earlier in the supply chain are very large in number, even fragmented with relatively less power. The need is to analyze whether partnership agreements exist between the companies and retailers are worth it or are they become a tool to force the manufacturers to pull down the prices and sell the stock to retailers which ultimately leads to their increase in profit margin. If so, companies are not in partnership and moved away from collaborative relationships. For the sake of establishing, a collaborative relationship among partners' processes or activities involving partnership needs to consider for evaluating the efficiency and effectiveness of textile supply chain (Bruce et al., 2004).

The ultimate product of the textile industry includes fibers, yarn, fabric, shirts, pants, bed sheets, towels, curtains, sweaters, etc. Configuration of the supply chain in textiles across international boundaries needs to be very right for high efficiency, effectiveness, and flexibility. Managing various activities need efficient strategies whereas, different processes need special solutions. The very first bottleneck in the textile industry is the lead time and factor of major uncertainty. The other two bottlenecking issues are fabric manufacturing, processing, and distribution. We are in need of strategies to facilitate the smooth working of these cyclic processes/activities. Efficient procurement strategies can make the industry enable to stock some raw materials; so as to fulfill the demand of weaving arrives. In accordance with the immediate customer, probable future orders/demands with specific requirements can be known to manufacturers. Continuing in the same manner, distribution centers can be stocked with a sufficient quantity of fabric, so that whenever fabric demands arrive, distribution centers can deploy their strategies to distribute the fabric at the preferred location. The location of the distribution center must be established closely, failing which, strategic ties should be established between fabric manufacturers and fabric processing or garment manufacturers. Efficient distribution channels play a vital role in executing delivery in minimum time with the economy. With this, opportunities must be sought to lower inventory cost, information carrying cost, reduce total supply chain cycle time, etc. Capabilities must be enhanced to achieve competitive advantage. The probable solution of the right strategies to issues such as procurement of raw material, production scheduling and flexibilities, and efficient distribution planning and scheduling can resolve the problems of bottleneck in textile supply chain management. If not, outsourcing entire processes would be a possible solution to gain significant flexibility in capacities, which in turn is not economical for a developing nation like India.

As stated earlier, quality control activities are most important barrier in attaining balance in supply chain with customer satisfaction. Determining and removing these barriers will give an efficient supply chain and will surely benefit the partners. The impact of carrying forward the quality defect may affect very seriously. If a defect is found at the end, that is, at the retailer and a situation may arise, that the whole lot is built with the wrong specification and yet another lot is also ready with the manufacturer, great financial loss may take place. There is a need to look to develop strategies that take into account the use of quality control aspects and may motivate to use quality engineering techniques and flexibility to satisfy customer needs.

We would like to move ahead with the conclusion that, the textile supply chain is very complex in nature and flow and it sounds very cumbersome to model and optimize every issue of the textile supply chain concurrently. The solution to such a complex condition is to develop a strategic perspective which takes care of such problems holistically. The textile supply chain has various research issues such as sourcing problems, quality control, flexibility in serving needs, forecasting of demands, production efficiency utilization, supply base, etc. Below is the list that summarizes some of the issues cited in the literature review and also emerged during the discussion with experts as shown in Table 2.2.

TABLE 2.2 Existing Problems in Textile Supply Chain with Associated Division/Processes.

Sr. No.	Problems/issues in textile supply chain	Associated division
1	Inefficient supplier selection	Sourcing
2	Weak purchasing/sourcing department	Sourcing
3	Poor coordination of suppliers and manufacturers due to weak information sharing	Sourcing
4	High cost of global sourcing.	Sourcing
5	Increased lead times and quality variation	Production
6	Inaccuracy of inventory holding system	Production
7	Quick changes in fashion	Production
8	Ignorance about mass customization	Production
9	Ignorance about mass customization	Production
10	Value-added products	Production
11	Loss of competitiveness	Production
12	Poor results of quick response	Distribution
13	Lack of marketing and management skills	Distribution
14	Incorrect location of distribution centers	Distribution

TABLE 2.2 *(Continued)*

Sr. No.	Problems/issues in textile supply chain	Associated division
15	Unsuccessful forecasting and manufacturing	Sourcing and Distribution
16	Short product cycle and forecasting errors	Sourcing and Distribution
17	Replenishment problems	Sourcing. Production and distribution
18	Limited implementation of vendor managed inventory (VMI)	Sourcing. Production and distribution
19	Selection of feasible ERP system solutions	Sourcing. Production and distribution
20	Poor IT setup and information invisibility	Sourcing. Production and distribution

As it is clear from the table, issues or problems in the textile supply chain are has been affecting almost all perspectives (internal operation, learning and development, and customer) of the textile supply chain which ultimately builds the financial burden on the industry. Hence, a comprehensive approach is needed to look at the textile supply chain to measure its performance.

Apart from this, some reports also suggest that the competitiveness of the supply chain is good during the initial stage of production of raw material, but it is reduced during the second stage of yarn and further to fabric production which mainly affects the various end of textile supply chain such garment manufacturer, bed furnishing production, etc. Also, it added that the spinning units are standalone and not involved in weaving – knitting – processing of the fabric, as a result of which they fail to integrate with the textile supply chain and also lack gaining profit advantage. Further, manufacturing of readymade garments or bed liners etc. are deeply affected by the de-synchronized behavior of spinning, weaving, knitting, and processing that creates gaps in the textile supply chain resulting in poor performance outcomes of the industry (Report by National Manufacturing Competitiveness Council, 2009).

It is also clear from the table that considering division independently, for example, sourcing or distribution will not solve the purpose of achieving efficient and effective supply chain management for the textile industry. Supply chain partners are associated very closely with each other via the processes of procurement, production, and distribution; thus, it is obvious that issues in the first division will surely affect progressive sections. For example, improper selection of suppliers or receipt of defective raw material affecting the procurement process will cause the production system to be more conscious of adding value to product manufacturing and make the distribution centers ship it in optimum time. This leads to affect more concentration on internal business processes for adding value, developing newer techniques, and learning environment to remain competitive hence, satisfying the customer needs and ultimately fulfilling the stakeholder expectation by earning satisfied profit gains.

It has been found by many researchers that information sharing among supply chain partners or between activities is one of the crucial factors for effective SCM. Stein and Sweat (1998) suggested that information exchange on a regular basis will facilitate supply chain partners to work effectively as a supply chain.

A hypothesis-based study has been conducted by Pratabaraj (2013) for the textile industry in India considered five dimensions of the supply chain which are quality, flexibility, dependability, innovation, and overall supply chain performance. The study tested four hypotheses, which deal with activities where the author stresses on procurement, production are the main activities, income groups dealing with companies' investment strength, age groups, and last overall supply chain performance. The study lacks in considering the issues related to distribution strategy, the cost associated with other activities, the efficiency of business processes, and the development of creative culture.

Prior to 2005, it has been cited in many studies that cost has been predominantly the major driver for supply chain effectiveness. However, the shift of research has taken the practitioner or even academician to look the supply chain efficiency in perspective, which deals with the non-financial aspect of the textile supply chain. These aspects are considering the measures focused on the satisfaction of customers, proper internal logistics which includes capacity utilization, inventory issues, etc., and last with which most textile industry would find it difficult to sustain a competitive global environment, that is, innovation and learning.

The ultimate aim or objective of the textile supply chain is to deliver the right textile products to the market for satisfying customers at the lowest possible cost, with great aesthetics at a fast speed by incorporating newer methods for serving the volatile demands of the market and gaining maximum profit. Also, the integrating process with the help of which the above objective would not be possible to achieve in the supply chain network having, procurement, production, and distribution as functional divisions. Hence, the most important aspect of the textile supply chain would be the finances involved, the efficiency of internal logistics, deployment level of innovation, and new learning environment, and finally degree of customer satisfaction.

2.5 SUMMARY

With the recognition of textile being an unorganized industry and obsolete in addressing the focus of researchers, and practitioners, and also with reference to the study of microcharacteristics of textile supply chain management, SCM has gained more attention in the textile industry by academician and researchers in the recent past. In the late 1980's with the advent of well-known initiative of quick response (QR) in the textile supply chain, some basic aspects gained more attention such as lead time results in evaluating the approach of QR. In association with QR, other issues such as logistics, production, and distribution also observed using different methodologies. This is achieved by effectively inspecting the issues related to the division of the supply chain network (procurement–production–distribution). Whereas, the literature does not provide the use of different divisions in developing any system for supply chain performance assessment in the textile industry.

Also, SCM is very critical in the textile industry, also, the study of the development of supply chain performance measurement is at a very nascent stage. Hence, dealing with the textile supply chain with very basic attributes of an industry would be the most fruitful justice to the industry and addition to state of art of supply chain management literature. As we all know, the cost has always remained one of the crucial perspectives of any industry, hence using the literature review other perspectives have also been focused on, which are provided as follows: Customer, internal business operation, and last innovation and learning. These four perspectives contribute to a balanced scorecard (BSC) as suggested by Kaplan and Nortan (1992) and are sufficient to give an overall foundation to the supply chain performance measurement model for the textile industry in this research.

KEYWORDS

- **supply chain practices**
- **supply chain optimization**
- **textile industry**
- **collaboration and coordination**

CHAPTER 3

SUPPLY CHAIN PERFORMANCE MEASUREMENT PRACTICES: CASE OF TEXTILE INDUSTRIES

ABSTRACT

Supply chain performance measurement (SCPM) plays a crucial role in the textile industry as it enables organizations to assess and enhance the efficiency and effectiveness of their SC operations. This abstract provides an overview of SCPM in the textile industry, focusing on the SCPM framework, a comparative analysis of studied models/frameworks, and performance measurement criteria. The SCPM framework serves as a structured approach to evaluate and monitor the performance of various elements within the SC, including suppliers, manufacturers, distributors, and retailers. It enables organizations to identify areas of improvement, optimize processes, and enhance overall SC performance. By implementing an SCPM framework, textile companies can gain valuable insights into key performance indicators (KPIs) such as cost, quality, delivery time, and customer satisfaction, allowing them to make data-driven decisions and align their SC strategies with business goals. A comparative analysis of studied models/frameworks offers a comprehensive assessment of different SCPM approaches employed in the textile industry. It enables organizations to understand the strengths and limitations of each model/framework and select the most suitable one based on their specific requirements and objectives. Performance measurement criteria for SCPM in the textile industry encompass a range of factors including on-time delivery,

Supply Chain Performance Measurement in Textile Enterprises. Pranav C. Charkha, PhD, Santosh B. Jaju, PhD & Prasenjit Chatterjee, PhD (Authors.)

order fulfillment rates, inventory turnover, product quality, lead time, and customer service levels. By effectively measuring these criteria, textile companies can identify bottlenecks, improve collaboration among SC partners, enhance visibility, and achieve operational excellence.

3.1 SCPM IN TEXTILE INDUSTRY

3.1.1 SUPPLY CHAIN PERFORMANCE MEASUREMENT (SCPM)

Performance measurement has gained immense importance in recent past and is still growing with its research opportunities. Various studies are being undertaken to review its techniques and methodologies for evaluating performance of business enterprises. Performance measurement system (PMS) has been evolved over the years. Owing to volatile changes taking place worldwide in all kinds of business enterprises, such as advent of new technologies, integration, coordination, many business enterprises have concentrated on the supply chain (SC) instead of focusing only on their inbound logistics. Hence, shift has taken place in organization from normal PMS to performance measurement of supply chains. It is explicitly brought out of literatures that comprehensive supply chain performance measurement (SCPM) is the need of today's competitive business environment (Aramyan et al.). Hence, it is very rightly said that supply chain performance assessment plays a vital role in SCM effectiveness and efficiency enhancement. Academicians, practitioners, and industry experts paying lot of attention toward measuring supply chain performance to improve the integration, coordination, and cooperation among supply chain partnering firms. Assessing supply chain performance acts as a bridge between expectation of stakeholders and execution, implementation of business practices, which ultimately enable enterprises to recognize prospective gray areas for improvement.

A performance metrics development model was proposed by Adams et al. (1995) containing 5 steps starting with business development strategy and ending with utilizing quality functional deployment (QFD) to develop performance metrics. Using this model, he proposed to develop critical success factors (CSFs) using QFD. CSFs are such goals which facilitate the attainment and execution of firm's strategy. The model suggested by Adams has positive aspects such as it aligns global CSFs to process CSFs and then links to measures. It utilizes transformational process approach which is elaborative, generic, and organized.

Model suggested by Adams et al. (1995), not necessarily be the actual performance measurement system, but presented an approach toward designing performance measures aligned with organizational strategy. Preferably suggested was not for supply chain, but can be deployed using certain modification for supply chain.

The most prominent model suggested by Kaplan and Nortan (1992, 1993, 1996) for designing performance measurement system translated business strategies and goals into objective and performance measures. They enveloped comprehensive nature of an enterprise involving perspectives like financial, customer, internal business, and innovation and learning.

Many studies have been conducted for manufacturing and service organization to develop the performance measurement system to measure performance. Process-oriented top-down method known as ENAPS was developed by Browne et al. (1998) having defined huge set of indicators. The strategically aligned measures were proposed by Hudson et al. (2001) being useful for continuous improvement. It is suggested to link performance measures to precise improvement initiatives, enabling attainment of strategic goals of organization, subjected to change, depending on scenario of market. Rouse et al. (2003) convicted with emphasis that developing performance measurement framework develops a culture of building performance measurement approaches by doing SWOT analysis. Folan and Browne (2005) developed the performance measurement system purposely for extended enterprise using two distinct approaches: balanced scorecard and practical tool for identifying and implementing measures.

Dixon et al. (1990) said that studies on performance measurement evolved periodically in two different phases; prior to the 1980s and later to the 1980s. The earlier period was concerned with assessing the performance of an industry based on the traditional cost aspects such as profit, capital investment, rate of return on investment, and labor productivity. On the other hand, later in the 1980s, phase focused on the development of quality management concepts such as total quality management, supply chain management, etc., with major stress on designing performance measures less tangible in nature with shift from financial to nonfinancial measures for performance evaluation.

It was suggested by Turner et al. (2005) that the application of performance measures since a decade is an evolutionary transformation from financial to nonfinancial measures. The reason behind the use of such nonfinancial metrics is just because of development of various approaches such as balanced scorecard (Kaplan and Nortan, 1992), ABCD checklist of operational excellence (Wight, 1993), and European excellence framework (developed by European foundation for quality Management, 2009).

The existing performance measurement system has limitations for manufacturing unit such as short-term concentration, focusing on minimizing errors and variance with respect to benchmark and oversight the process continual improvement lying in results and determinants as well internal and external focus, absence on external focus (Kaplan and Nortan, 1992) and dealing with multiple perspective of stakeholder (Kennerley and Neely, 2002)

The best perspective for the performance measurement system was suggested by Neely et al. (1995), which has become a common base for recent research on the existing supply chain performance measurement system for developing performance measures/metrics (as suggested by Beamon et al., 1999, 2001; Bourne et al., 2003, Gunasekaran et al., 2001, 2004, 2007). Many more studies have addressed the limitation cited by Neely et al. (1999), which becomes basis for the development performance measurement for supply chain. Besides, they stressed the development of performance measurement system for supply chain overcoming addressed limitation. It is proposed to use any of the earlier approaches, subject to fulfillment of the exact expectation of supply chain to measure the performance. These suggestions and proposals are step forward and with the help of this study, it is intended to stress upon the development of a suitable performance measurement system considering existing practical issues in textile industry in developing countries like India.

3.2 SUPPLY CHAIN PERFORMANCE MEASUREMENT FRAMEWORK

Some standard model/framework must be used to measure the performance of supply chain management. Various studies were conducted to demonstrate the application of different types of performance measures used for assessing supply chain management, such as Beamon (1999) explained the use of three types of performance measures namely measures (resources, output, and flexibility) for manufacturing supply chains. Yet, these performance measures are important for any performance measurement system and a flexibility quantitative measurement approach for supply chains is proposed, but it has limitation in considering the system perspective, in which a performance of the supply chain must be measured as whole.

Another study done by Gunasekaran et al. (2001) suggested various performance measures and metrics for evaluating supply chain management using a framework which gives us what to measure, and how to do task. The framework categorized performance measures at three levels of decision-making as strategic, tactical, and operational levels of management, as well as divided into financial and nonfinancial. The disadvantage of this framework is the huge number of performance measures. This poses difficulty for firms that which to use and when, as no firm will use all measures in day-to-day operation. Again, it has not provided any methodology for the prioritization of these measures and no consideration of overall perspective is seen in the framework.

An innovative performance measurement system was suggested by Chan and Qi (2003) adding to the progress of supply chain management from five processes: supplying, inbound logistics, core manufacturing, outbound logistics, and marketing and sales. These processes are used to measure the holistic performance of intricate supply chains. A real-world situation is tested by using the fuzzy set theory method. The proposed framework lacks in considering again overall perspective and various decision-making aspects at the strategic, tactical, and operational levels.

An attempt was made by Otto and Kotza (2003) to design metrics to measure the effectiveness of supply chain management, categorized under six supply chain perspectives which are system dynamics, operations research or information technology, logistics, marketing, organization, and strategy. Every perspective has its own idea for supply chain, issues and solutions, and associated performance metrics. But, the limitation to the study is not every metric is used by enterprise and again, selection of perspective should be based on the concept of involvement of every stakeholder.

Yet, another study by Gunasekaran et al. (2004) developed a framework to convey the significance of supply chain management performance measurement and corresponding metrics. The framework suggested uses the supply chain operation reference (SCOR) model processes (plan, source, make, and deliver) in consideration with strategic, tactical, and operational decision-making levels. It used three-level importance criteria to prioritize metrics and conducted empirical analysis of few British companies. It has dearth in suggesting critical success factors supply chain management. Also, if its processes are considered, such as plan, source, make, and deliver, then it becomes more important to validate the score with the help of supply chain partners interacting for those processes like suppliers and customers to suggest possible improvement in supply chain performance.

Following work on SCOR, Huang et al. (2005) presented a summary on the supply chain operations reference model, its pros and cons with descriptive case studies and suggested a sophisticated tool to build up a supply chain threaded diagram as per SCOR requirements. Designing the configuration of supply chain in a special feature of SCOR implementation, in existence, the configuration of "as-is" or "to-be" threaded-diagram is used. A high-end sophisticated tool has been suggested and explained in this paper for deciding configuration. The proposed method can only consider single manufacturing facility of a company and not the processes across supply chain partners.

Aramyan et al. (2007) suggested the framework Dutch--German tomato supply chain performance measurement. It is concluded that for tomato supply chain performance measurement four performance measures are vital and they are efficiency, flexibility, responsiveness, and food quality. They also proposed an integrated performance measurement system that used both financial and nonfinancial indicators mapped with features of food supply chains. As the study is conducted in one particular company from the Dutch--German tomato supply chain, results cannot be generalized and hence can be limitation to work.

Also, Bhagwat and Sharma (2007a) proposed the use of balanced scorecard (BSC) for measuring performance of supply chain management using the following four perspectives: finance, customer, internal business process, and learning and growth. Three small- and medium-scale enterprises in India were taken for conducting the study. The developed model provides a useful guidance for the managers in assessing and measuring of supply chain management. However, a further proposal is suggested to establish whether the projected aspects are sufficient and adequate; further studies were conducted in independent industry and did not consider the supply chain comprising multiple firms.

Jammernegg and Reiner (2007) conferred the improvement opportunities and difficulties in the performance measurement of supply chain processes by the synchronized use of inventory management with capacity management. A simulation-based model is used for the selection of a supplier in telecommunication and automotive industry. The simulation model shows the effect of coordination between inventory management and capacity management results in performance improvement. This study is conducted to review the cost and service aspect of an industry, hence, to view the whole supply chain process.

Yeh et al. (2007) developed a model to evaluate the performance of supply chain using fuzzy linguistic computing (FLC). The Six Sigma: define, measure, analyze, improve, and control (DMAIC) method is utilized in designing the framework. The Fuzzy approach is modified to make available the aggregation algorithm to ensure consistency. Earlier, the Delphi method is utilized to seek the responses from subject matter experts for finalizing criterion, performance measurement factors, etc. This framework lacks in considering the decision-making levels and supply chain perspective.

Study on sourcing processes and strategy and associated metrics for performance using SCOR model and regression analysis. The outcomes were again reviewed for sourcing process of level 3. A questionnaire survey method is used to gather observed information from the monitor display manufacturing industry in Taiwan. The study also presented the implementation process of SCOR and examined the project planning concept using the SCOR model. Due modification and revision would be needed as the study uses SCOR 7.0 only.

Robb et al. (2008) developed a model to study the relationship among supply chain and financial performance by using a structural equation model (SEM) with China furniture manufacturers of study. The study is conducted in Chinese furniture manufacturing which are facing issues pertaining to low labor productivity, low exports affecting the substantial growth. The study highlighted comparative importance of supply chain and operational practices and concluded that operating attributes affect the business performance. The other conclusion of the study is that human resource and innovation is a strongest link; hence, inhibiting leaning environment is important in gaining a competitive advantage. The study is limited to considering only operational dimension, not the entire supply chains.

A more user-friendly method is proposed for the model of Chan and Qi (2003) by Theeranuphattana and Tang (2008) using the fuzzy logic technique. The suggested method is a combination of Chan and Qi's model and (SCOR) model. These two methods complement each other when measuring supply chain performance. To find the application of the proposed method, one case is presented from the Thai manufacturing company.

In a study among Chinese manufactures, Zhu et al. (2008) attempted to develop a model applied to assess the practical implementation of the green supply chain management. The scale is developed using the survey questionnaire designed from the literature available and opinions of experts from academics and practitioners. Factor analysis is used to compare the parameters of two models developed for measuring practices of GSCM implementation. The focus of the study is on environmental performance assessment, and did not include operational performance and hence excluded whole supply chain performance.

Cai et al. (2009) suggested a framework for managing the performance of the supply chain in a dynamic environment. It systematically and quantitatively analyzes interdependent behavior among the performance measures using KPI accomplishment cost transformation matrix (PCTM). The proposed method is an extension of the traditional eigen value method and work transformation matrix (WTM) evolved from decision structure matrix (DSM). The results obtained from the proposed method cannot be readily adopted as final decisions, rather can be supplementary information for taking decision.

Chae (2009) presented the practical method to evaluate the performance of supply chain management. The study provided a list of performance measures ready to use for various enterprises. The framework is developed based on the SCOR level 1 metrics. It is found from the responses of the industry that less number of metrics is beneficial. Hence, enterprises must consider a limited number of key performance indicators important for their organizational perspectives. The framework lacks in considering "return" process of the SCOR model.

Chia et al. (2009) used balanced scorecard (BSC) for measuring supply chain performance of logistic industry. The proposed study elaborates what executive of industry measures and how they deal with the perspective of performance measurement using BSC. The questionnaire survey method is utilized to seek the opinion of experts from the logistics industry. The outcome of the study is lag in strategic coordination with performance measurement of internal business processes and learning and growth performance criteria. It also added that BSC can be viewed as a very comprehensive tool to evaluate the supply chain performance. The two perspectives of internal business, innovation and learning, are not measured as they are drivers of sustainable growth. The targeted respondent cluster is relatively smaller in size; hence, results cannot be representative of an independent cluster. Also, it lacks accountability of alternatives using which BSC would measure the performance.

Rodriguez et al. (2009) developed quantitative relationship performance measurement system (QRPMS). The model suggested presents a unique approach to identify and quantify relation among performance measures selected within a framework for performance measurement based on balanced scorecard (BSC). Use of BSC is to facilitate supplementary information to managers to take decisions. The model is applied to a clothing manufacturer of Spain. The limitation to the study is as this model is tested in one industry only, it limits the generalization of results.

An attempt was made in India for small- and medium-scale enterprises by Thakkar et al. (2009) who suggested an integrated supply chain performance measurement framework for the supply chain using a set of qualitative and quantitative parameters studied during a case study conducted in the said industry. The framework suggested utilized the features of two renowned predefined models: balanced scorecard (BSC) and supply chain operation reference (SCOR) model. The study also presented an implementation strategy of the framework. The study coordinated measures with cycles like procurement, manufacturing, replenishment, and customer order. But it has not considered distribution and different levels of decision-making.

Bigliardi and Bottani (2010) found the application of balanced scorecard (BSC) model in the food supply chain performance measurement. The developed model can be treated as reference to establish key performance indicators for the food industry. The study consists of an extensive literature survey pertaining to performance measurement framework and measures, existing performance measurement systems in food industry and application of BSC model, and associated financial and nonfinancial indicators, relevant to use. The Delphi technique is employed for the finalization of framework and performance metrics. For validation, the final model is examined in two food companies. But, as a matter of fact that selected field of application of model is very different in nature of supply chain which possibly can be the limitation of the study as the result cannot be applied to other contexts.

Flynn et al. (2010) analyzed approaches of contingency and a configuration perspective. They used these perspectives to establish a relationship among supply chain integration dimensions, operational and business performance from the above two perspectives. In this study, supply chain integration is defined as of level manufacturer to strategically coordinate with its partners and controls processes both within and across the supply chain so as to accomplish effective and efficient product flows and services, information, money, and decisions, in order to offer utmost importance to the customer. While using a contingency approach, the hierarchical regression method is applied to decide the impact of individual integrating partners (supplier, internal, and customer integration) and their interactions on performance. But, while using the configuration approach, the cluster analysis technique is used to develop patterns of supply chain integration. To examine the relation between supply chain integration pattern and performance, analysis of variance is used. The study lacks in considering decision-making level and also it lacks in focusing on various perspectives of an organization.

Lin et al. (2010) focused on factor of innovation in supply chain integration. The study was conducted by Taiwan-qualified high-tech manufacturers. The research study was based on a literature review of earlier work on strategic orientation, innovative practices in channel integration, and its effect on performance to develop a theoretical framework and hypotheses. The study uses confirmatory factor analysis (CFA) through AMOS 7.0 analysis. The study shows that a considerable relationship exists between market orientation supply chain performances. But, this model only focused on the innovation perspective.

The taxonomy of earlier work on supply chain performance measurement is presented in Table 3.1.

Pratabaraj et al. (2013) conducted a study in India in the textile industry to investigate the performance of the textile supply chain. The study used an approach of hypothesis testing. It used only four industry-centric performance metrics to assess the performance. Quality, Flexibility, Dependability, and Innovation are the performance metrics. The study has not considered financial, customer, and business functions. The limitation to the study is that it focuses on the industry involved in all sorts of operation under one roof from yarn to garment manufacturing. This isolates the essence of supply chain and integrates various partners in it.

After reviewing literature, it is understood that designing any performance measurement system is not a task of choosing the crucial measure/metrics only, but rather is the system of acknowledging and attaining overall consensus in determining and developing appropriate performance measures/metrics irrespective of the type of industry and supply chain.

Although numerous research studies have been done in terms of performance measurement of supply chains, textile industry concentrated research is still at its nascent stage. Very limited numbers of efforts were done for the textile industry (Lohman et al., 2004; McMullen, 1996; Nowell, 2005; Sauls, 2007).

The taxonomy of literature reviewed with respect to the supply performance measurement framework has let us understand that most of the studies have frequently paid attention to considering single area of performance measurement (Beamon, 1999; Gunasekaran et al., 2001; Chan and Qi, 2003; Otto and Kotza, 2003; Huang et al., 2005; Aramyan et al., 2007; Berrah and Cliville, 2007; Jammernegg and Reiner, 2007; Yeh et al., 2007; Zhu et al., 2008; Chae, 2009; Chia et al., 2009; Lin et al., 2010), whereas research on a collective approach for the entire performance of supply chain management has been reasonably deserted. Even the studies, which were undertaken, were mostly done in developed countries. Very few were concentrated on developing countries. Also the industries which were taken up for application or validation are preferably electronics industry, food industry, or logistics industry. The consideration of the textile industry is at a very nascent stage. Pertaining to these gaps, this research study, therefore, is anticipated to offer a comprehensive approach to look at the aspect of supply chain performance measurement framework by using both supply chain cyclic processes (procurement, production, and distribution) and balance scorecard (BSC) perspectives and thereby adopting a cross organizational approach.

The above-mentioned approaches are utilized in research of supply chain management in general, but they have not been cited and applied collectively before to address the issue of performance measurement framework in supply chain management of textile countries in developing nations. The proposed framework facilitates the textile enterprises of developing countries to understand the overall perspective on supply chain performance. The framework consists of a balanced

TABLE 3.1 Taxonomy of Performance Measurement Frameworks Developed for Supply Chain Management.

Sr. No.	Authors	Framework/ measurement area	Remarks (strength (S) and weaknesses (W))	Geographical location	Application industry
1	Beamon (1999)	Supply chain processes	**S:** Identify three types of performance measures and propose flexibility quantitative measurement approach for supply chains. **W:** Lack of system thinking of measuring supply chain widely across the whole.	United States	Discrete Manufacturing Industry
2	Gunasekaran et al. (2001)	Decision-making levels	**S:** Combine decision-making levels with financial and nonfinancial criteria **W** Too many numbers of metrics and measures.	United States	Not specified
3	Chan and Qi (2003)	Supply chain processes	**S:** Identify five core processes as holistic complex supply chain measurement and introduce fuzzy set theory for judgment and evaluation processes. **W:** Overlook on the decision-making ability.	Hong Kong	Not specified
4	Otto and Kotza (2003)	Supply chain perspectives	**S:** Design six unique sets of supply chain metrics to measure the effectiveness of supply chain management. **W:** All the metrics are not used in business practice to measure supply chain performance.	Germany	Not specified

TABLE 3.1 Taxonomy

Sr. No.	Authors	Framework/ measurement area	Remarks (strength (S) and weaknesses (W))	Geographical location	Application industry
5	Gunasekaran et al. (2004)	Decision-making levels	**S:** Consider supply chain processes with respect to decision-making levels. **W:** Need collaboration from all stakeholders of supply chain system to evaluate the framework.	United Kingdom	Not specified— large scale industries
6	Huang et al. (2005)	Supply chain operations reference (SCOR) model	**S:** Summarize SCOR model and describe a computer-assisted tool to configure supply chain. **W:** Limit to one manufacturing company for studying.	United States	Manufacturing company
7	Aramyan et al. (2007)	Financial versus nonfinancial	**S:** Develop an integrated performance measurement system that contains financial as well as nonfinancial indicators. **W:** Investigate only single-food company.	The Netherlands	Agri-food supply chain
8	Berrah and Cliville (2007)	SCOR model	**S:** Use performance indicators from Gunasekaran et al. (2004) and apply MACBETH methodology to the supply chain processes. **W:** Does not consider the return process.	France	Not specified

TABLE 3.1 Taxonomy

Sr. No.	Authors	Framework/ measurement area	Remarks (strength (S) and weaknesses (W))	Geographical location	Application industry
9	Bhagwat and Sharma (2007a)	Balanced score card perspective	**S:** Develop a balanced scorecard for supply chain management and apply in SMEs in India. **W:** Need to determine the proposed perspectives and measures.	India	Not specified— SMEs
10	Jammernegg and Reiner (2007)	Intraorganizational versus interorganizational	**S:** Deal with performance measurement and improvement of supply chain processes by coordinated application of inventory management and capacity management. **W:** Lack to view the whole supply chain process as concentrate only on costs and service level.	Switzerland	Telecommunication and automotive industry
11	Yeh et al. (2007)	Six Sigma (DMAIC) processes	**S:** Propose a modified 2-tuple fuzzy linguistic computing (FLC) model to evaluate the performance of supply chain management. **W:** Lack of combining the decision-making levels.	Taiwan	Discrete manufacturing industry
12	Hwang et al. (2008)	SCOR model	**S:** Aim to find out important SCOR sourcing performance metrics by using the stepwise regression model. **W:** Concentrate on SCOR model version 7 only as version 11 will launch soon this year.	Taiwan	TFT-LCD Industry

TABLE 3.1 Taxonomy

Sr. No.	Authors	Framework/ measurement area	Remarks (strength (S) and weaknesses (W))	Geographical location	Application industry
12	Robb et al. (2008)	Operations practice and performance	**S:** Propose a model exploring operations practice and performance of supply chain management. **W:** Study only operations dimension, not for the whole supply chains.	China	Furniture manufacturing
13	Theeranuphattana and Tang (2008)	Combined Chan and Qi (2003) model and SCOR model	**S:** Develop a practical and efficient measurement model by incorporating the strengths of two different models. **W:** Present only one Thai manufacturing company case study.	Hong Kong	Not specified (case supply chain)
14	Zhu et al. (2008)	Green supply chain management	**S:** Investigate green supply chain practices implementation. **W:** Concentrate only on environmental performance with operational performance.	China	Chinese manufacturers
15	Cai et al. (2009)	Supply chain processes	**S:** Propose a framework using a systematic approach to improving the KPIs in a supply chain context. **W:** Result from PCTM analysis can use as only supporting information for making decision, not as direct decision.	China	Grocery Retail Company

TABLE 3.1 Taxonomy

Sr. No.	Authors	Framework/ measurement area	Remarks (strength (S) and weaknesses (W))	Geographical location	Application industry
16	Chae (2009)	SCOR model	**S:** Offer a practical approach to performance measurement and propose key performance metrics. **W:** The return process is not considered in this work.	United States	Nonspecified
17	Chia et al. (2009)	Balanced score card perspective	**S:** Apply the balanced scorecard approach on the logistics industry to measure performance of supply chain. **W:** The sizes of respondent clusters are different.	Singapore	Logistic industry
18	Rodriguez et al. (2009)	Balanced score card perspective	**S:** Propose the quantitative relationships performance measurement system based on the balanced scorecard. **W:** Study only one manufacturing company.	Spain	SME
19	Thakkar et al. (2009)	Integrated balanced score card and SCOR model	**S:** Develop an integrated performance measurement framework for supply chain evaluation and planning in SMEs. **W:** Do not consider decision-making levels.	India	Not specified- SMEs
20	Bigliardi and Bottani (2010)	Balanced score card perspective	**S:** Develop a balanced scorecard model for measuring performance in the food supply chain. **W:** Examine only specific industry field (the food industry).	Italy	Discrete manufacturing company

TABLE 3.1 *(Continued)*

Sr. No.	Authors	Framework/ measurement area	Remarks (strength (S) and weaknesses (W))	Geographical location	Application industry
21	Flynn et al. (2010)	Supply chain integration, operational, and business performance	**S:** Study the relationship between three dimensions of supply chain integration, operational and business performance. **W:** Overlook the decision-making ability.	United States	Not specified
22	Lin et al. (2010)	Supply chain innovation	**S:** Propose a model of innovation drivers in channel integration in supply chain management. **W:** Concentrate on innovation perspective only.	Taiwan	High-tech industries

Source: Reprinted with permission from Pasutham, 2020. Copyright Dr. Asawin Pasutham scorecard perspective to evaluate the performance of supply chain cyclic processes, that is, procurement, production, and distribution. Each of four BSC perspective, financial, customer, internal business, innovation and learning, will have a specific list of performance measures categorized from the study of Bhagwat et al. (2007a) under three decision-making levels.

3.3 SUPPLY CHAIN PERFORMANCE MEASUREMENT MODELS

Very rare attempts have been made to precisely gather models for supply chain performance evaluation. Furthermore, there exists a debate on categorizing approach also. Following is the categorization of performance measurement models based on diverse criteria as classified by various researchers (Gunasekaran et al., 2007);

- Balanced scorecard perspective
  (financial, customer, internal business process, and learning and growth)
- Constituents of performance metrics
  (resource, output, flexibility)
- Supply chain operation reference model—place of measures in supply chain network
  (plan, source, make, and deliver)
- Decision-making levels
  (strategic, tactical, and operational)
- Character of measures
  (financial and nonfinancial)
- Base of assessment
  (quantitative and nonquantitative)
- Conventional versus modern measures
  (function-based and value-based)

To summarize, based on an intense review of literature of various studies, the chapter presents a structured categorization of models and approaches that are recognized as an important contribution toward the supply chain performance measurement. Framework of studies mentioned earlier is classified as three main models utilized incessantly in various research studies as shown in Table 3.2.

TABLE 3.2 Categories of Supply Chain Performance Measurement Models.

Sr. No.	Approaches	Criteria	Contributor
1	Balanced score card (BSC)	Outcome based	Kaplan and Norton (1992, 1996, 1997)
2	Supply chain operations reference (SCOR) model	Activity based	Supply Chain Council (1996)
3	Decision-making levels	Hierarchical	Gunasekaran et al. (2001, 2004)

Source: Reprinted with permission from Pasutham, 2012. Copyright Dr. Asawin Pasutham Supply chain performance measurement framework—case studies on the Thai manufactures.

3.3.1 OUTCOME BASED

The outcome/output-based model assembles both common performance parameters as well as cause and effect assumptions determining the collaboration among performance attributes. The most pervasive performance measurement technique in this is the balanced scorecard (BSC). The approach of BSC is pioneered by David Norton, the CEO of Nolan Norton Institute, and Robert Kaplan, a professor at Harvard University. It is used to evaluate enterprise's output from four diverse, yet comprehensive, perspectives, consisting of financial, customer, internal business process, and learning and growth, associating them with four stakeholders of an organization top management, customer, employees, and associated vendors/suppliers (Kaplan and Norton, 1992). The BSC criteria include both financial and nonfinancial measures from all four perspectives. The BSC approach facilitates top management to monitor organization's performance with multi-facet focus using a global information system. It also empowers managers with a tool which would be beneficial to explore opportunities for gaining a competitive advantage (Kaplan and Norton, 1996). The utmost important philosophy of BSC is the inclusion of nonfinancial measures along with standard financial measures. It has a wide application in framing and developing the comprehensive and realistic strategic framework consisting of aim and objectives to gain the competitive success (Kaplan and Norton, 1997). The identification, selection, and categorization of useful performance measures under BSC perspective has been done by many research studies (Bhagwat and Sharma, 2007a; Chia et al., 2009; Rodriguez et al., 2009; Thakkar et al., 2009; Bigliardi and Bottani, 2010).

3.3.2 ACTIVITY-BASED MODEL

To have a proper selection of performance measures, it becomes mandatory to clearly identify the activities and important processes in supply chain management. Involving and integrating all the activities from supplier to customer is very decisive for effective implementation of supply chain management. As the concept of supply chain is multifaceted, the need has been felt by various studies to develop new methodologies that can consider the performance of important activities/processes in the supply chain. A well-known and efficient model for the activity/process-based approach is the supply chain operations reference (SCOR). It came into existence in 1996 and was authorized by the Supply Chain Council (SCC), a leading organization of global firms working in the areas of supply chain management. It is coined as a detailed "business process framework" which combines performance measures associated best practices, and software requirements (The Supply Chain Council, 2009a). The extent of the SCOR model involves all parameters for satisfying customer's demand initiating with demand forecasting and ending with demand satisfaction. As stated earlier, it is a business process reference model, having a structure that includes business processes, measures, best practices, and technology characteristics. It integrates philosophy of business process reengineering (BPR), setting standards, process evaluation, and implementation of best practice for achieving effective and efficient supply chains. Five distinct management activities, namely, plan, source, make, deliver, and return made the original SCOR model. In addition to this, a variety of performance measures are associated with these processes allied with five performances attributes: reliability, responsiveness, flexibility, cost, and asset metrics (The Supply Chain Council, 2009b). Studies in the field of supply chain management involve various applications of the SCOR model for the selection of appropriate performance measures (Huang et al., 2005; Berrah and Cliville, 2007; Hwang et al., 2008; Theeranuphattana and Tang, 2008; Chae, 2009; Thakkar et al., 2009).

3.3.3 HIERARCHY-BASED MODEL

This model of performance evaluation, monitor, and measures performance uses different hierarchical levels of an enterprise. The purpose of reviewing issues of supply chain performance measures at strategic, tactical, and operational levels is to facilitate the organization to take an appropriate decision in order to reach to the aim and strategic objectives. The strategy's success depends upon its transformation into operations. Gunasekaran et al. (2001) categorized performance measures accordingly with respect to decision level prospects: strategic, tactical, and operational. The complex behavior of supply chain management motivates to design a list of most effective key performance indicators at each hierarchical level. Utilization of appropriate measures at each hierarchy ascertains the achievement of objectives at that hierarchical level. Strategic hierarchical level favors intangible financial indicators (Gunasekaran and Kobu, 2007), also Gunasekaran et al. (2004) developed a performance measurement system for supply chain management. The proposed model integrates SCOR activities (e.g., plan, source, make, and deliver) with strategic, tactical, and operational levels and prioritizes each metric by three levels of importance: highly, moderately, and less important. Work is done by empirically analyzing the British company.

3.4 COMPARATIVE ANALYSIS OF STUDIED MODEL/ FRAMEWORK

It is very clear that the advent discussed in the earlier section has their diverse and specific focus in general. The effect of the outcome-based model is to explain the effect of management of internal factors which influences desired external outcomes. Hence, it is grouped under the outcome-oriented approach. The comparative advantage and limitation of these models are listed in Table 3.3.

The renowned approach is balanced scorecard (BSC) suggested by Kaplan and Norton (1992). BSC is a very influential, comprehensive strategic management tool that facilitates top management to design the strategy, implementing it utilizing proper measures to attain the corporate vision, mission, and objectives (Kaplan and Norton, 1996).

In this way, it enables senior executives/managers to develop and regulate the vision and strategy of the organization, keeping concentration on crucial performance indicators for existing and future business processes (Chia et al., 2009). A symmetrical assessment of supply chain performance using balanced scorecard allows organizations to monitor the progress of operation with faster and wider pace and also assists firm to initiate improvement in business functions such as product design applications, production facilities, quality improvement and benchmarking, inventory management, reduction in lead time, attainment reputation in market shares and effectively implementing business strategies (Bhagwat and Sharma, 2007a). Besides, few more benefits company may attain are mentioned in literature as written below (Bremser and Barsky 2004):

- Unambiguity in translating firm's strategy into operational terms,
- Configuring the strategy with organization structure,
- Converting strategy into a continuous process,
- Process accomplishment for learning and adapting the company's strategy,
- Creating leaders to drive change for strategic management system.

The second in the classification is an activity-based approach. Supply chain operations reference (SCOR) model is most preferred in this group. This is a widely used supply chain model by various industries. With SCOR, philosophical approach for all the supply chain partners becomes unique and they prefer to use uniform definitions for processes/activities, elements, and associated metrics. As the explanation for performance measures is very straightforward and standardized, the task becomes simple to use applicable measures. Proportion of using SCOR model by industries for supply chain performance evaluation has increased (Theeranuphattana and Tang, 2008). It facilitates the enterprise to assess the performance of supply chain in a structured manner, which improves information sharing among supply chain partner to design the most effective network of supply chain (Hwang et al., 2008). Strength of SCOR as notified by Thakkar et al. (2009) is that it enhances communication. It serves top management to structure the supply chain to achieve a desired performance (Huan et al., 2004). Instead of being its advantages, some studies have mentioned that even though the SCOR is activity and efficiency oriented, but is not strategy oriented (Gammelgaard and Vesth, 2004; Angerhofer and Angelides, 2006). It does not also allow organized prioritization of performance measures. Still one attempt has been made to couple it with analytic hierarchy processing (AHP) for decision-making (Huan et al., 2004). Besides this, it lacks in establishing coordination among functional strategies and corporate strategies which may affect resource alignment in the organization. Organization's management wishing to pursue the implementation of SCOR wanted that the framework should envelop entire strategy of enterprise, which helps them to synchronize the goals, objective with resources and achieve the result with its implementation (Lambert et al., 2005).

The last category of the group consists of hierarchical approaches of supply chain performance measurement. Prime contribution to this group is model proposed by Gunasekaran et al. (2001). In the last decade, enterprises have recognized the importance of supply chain management in their routine operations. But, they still lack in selecting the right and effective performance measures for supply chain management, which may be because of the absence of a balanced approach and ambiguity in differentiating metrics at strategic, tactical, and operational levels (Hudson et al., 2001). It becomes mandatory for achieving an efficient and effective supply chain that organizational strategic aims and objective must ensure the proper selection of measures/metrics should also be stratified as strategic, tactical, and operational levels with both attributes as financial and nonfinancial. The gap was filled by Gunasekaran et al. (2001, 2004) providing performance measures/metrics at strategic, tactical, and operational levels. Decision-making levels have been greatly affected by the selection of a performance measurement system. As a result of which, Gunasekaran et al. (2001, 2004) proposed frameworks that relate strategies to operational measures. Such a model enables managers to formulate and regulate strategy and percolate them down the hierarchy. This model implicates the integration of vertical levels and does not envelop the horizontal approach. Hence, such a model needs to couple with any other holistic model to cover the perspective of overall supply chain.

TABLE 3.3 Benefit and Constraint of Model Reviewed.

Sr. No.	Models/ approaches	Authors	Benefits	Constraints
1.	Balanced scorecard	Kaplan and Nortan	Balance of financial measures with nonfinancial measures. Motivates formulation and regulation consistent strategy in the organization Takes care of all stakeholders.	Lacks in integrating the measures at strategic level and operational level.
2.	Supply chain operation reference model	Supply chain council	A well-recognized and widely used as a standard for measuring the performance. Provide a systematic way to analyze supply chain performance.	Lack of systematically prioritizing measures. Also lacks in considering stakeholder aspect.
3.	Hierarchical model	Gunasekaran	Link the company's strategic plan down to operational level. Support decision makers to make the right decisions on each level.	Restricted to single organization only. Need to couple with other frameworks to focus on the whole aspect of supply chain.

Lot of criticism has been made by various authors on the modern studies on performance measurement frameworks for supply chain management (Holmberg, 2000; van Hoek, 2001; Hudson et al., 2001). This criticism has few valid issues of concentration which are: correlating with strategy, absence of system orientation, lagging balanced approach to incorporate financial and nonfinancial measures, and mostly not considering supply chain context.

Most of the studies (Simatupang and Sridharan, 2005; Angerhofer and Angelides, 2006; Busi and Bititci, 2006) suggested that consideration of supply chain cyclic processes as a success measure for organizations, but they lack in integrating these with respect to the focus of stakeholder consideration so as to look at the perspective of supply chain as a whole and measure it. Furthermore, a lot of studies were conducted to propose a number of theoretical models for performance measurements of supply chain management. But they lack in having observed analysis with implication in companies for performance measures/metrics for supply chain management, preferably for developing nations like India (Gunasekaran et al., 2004). Also, the studies lack in differentiating approaches of supply chain performance measurement for discrete manufacturing and process industry. As the supply chain for the process industry is very complex, it has to be dealt with differently (Pratabaraj (2013)). Hence, the proposed study wishes to link such laggings by suggesting performance measurement model for textile supply chain management which dispenses a concrete balanced approach. The suggested conceptual model will integrate supply chain cyclic processes from procurement, production, and distribution with four perspectives of balanced scorecard (financial, customer, internal business operations, innovation and learning). The subcriteria under each of the BSC perspectives would be consented by putting the list of performance measures/metrics proposed by Gunasekaran (2004) as utilized by Bhagwat et al. (2007a).

3.5 PERFORMANCE MEASUREMENT CRITERIA FOR SUPPLY CHAIN PERFORMANCE MEASUREMENT

For the success of an organization's existence, the feedback mechanism becomes the most important system. The driver to efficient supply chain management is the feedback in the form of functionality of performance measures/metrics (Chae, 2009). The supply chain performance measurement enables companies to efficiently plan and execute the strategies, in order to identify potential areas of improvement. It is being realized by a number of studies that identification of proper performance measurement criteria is not a plain sailing task. It is because no structured procedure is available for designing the supply chain performance measurement criteria for companies.

As per the practices reviewed from the literature of Table 3.3 (frameworks of supply chain performance measurement), various performance measurement criteria for supply chain management are listed in Table 3.4 as mentioned below. The term performance measurement criterion relates to an activity/process to be monitored for achieving organizational mission and objectives (Beamon, 1999). For reference, thus, companies may select the performance measurement criteria for evaluating the supply chain performance.

Supply chain management has been key driving philosophy for a number of enterprises in their routine business processes. On the other hand, they frequently lag in designing effective performance measurement criteria that are to be employed to achieve integrated supply chain management. It is just because of not clearly considering the stakeholder aspect, especially for textile supply chain cyclic processes that are the coordinating processes of supply chain partners. It is also because of not clearly distinguishing the criteria at strategic, tactical, and operational levels (Kaplan and Notan, 1996, Gunasekaran et al., 2001). Hence, it becomes important that for achieving the efficient and effective supply chain management, selection of proper performance measurement criteria must be done keeping the stakeholder's aspect in focus. Also the selection of overall supply chain perspective should be done by considering the supply chain cyclic processes. Hence, the design of effective supply chain performance measurement system must consider the strategic goals of an organization relating to different perspectives such as profit attainment, customer satisfaction, deploying the innovative and learning environment to get all the internal business operation functioning properly. Also, it should envelop complete processes of supply chain as a whole.

Designing the effective supply chain performance measurement system with efficient performance measurement criteria is a complex process and cumbersome for local business enterprises (Chan and Qi, 2003; Aramyan et al., 2007). In general, enterprises rely on performance measures such as cost, return on investment which evaluates firm's financial performance, whereas performance measures related to supply chain have remained uninformed to industries in literature. Lack of practical guidelines is problem for nowaday's enterprises (Lapide, 2000). As a result of which, this study proposed to connect these spaces by designing performance measurement system for supply chain management consisting of performance measurement factor categorized by Gunasekaran (2001) and

TABLE 3.4 Summary of Performance Measurement Criteria for Supply Chain Performance Evaluation.

Sr. No.	Authors	Measurement perspective	Performance measurement criteria
1	Beamon (1999)	Resources	Total cost of resources used – Total cost of distribution including transportation and handling costs. – Total cost of manufacturing including labor, maintenance, and rework costs. – Inventory costs associated with held inventory: inventory investment, inventory obsolescence, work-in-process inventories, finished goods inventories. – Return on investment (ROI).
		Output	Sales – Profit – Fill rate: target fill rate achievement, average item fill rate – On-time deliveries: product lateness, average lateness of orders, average earliness of orders, percent on-time deliveries Backorder or stock-out: stock-out probability, number of backorders, number of stock-outs, average backorder – Customer response time – Manufacturing lead time – Shipping errors – Customer complaints.

TABLE 3.4 *(Continued)*

Sr. No.	Authors	Measurement perspective	Performance measurement criteria
		Flexibility	Reductions in the number of backorders – Reductions in the number of lost sales – Reductions in the number of late orders – Increased customer satisfaction – Ability to respond to and accommodate demand variations – Ability to respond to and accommodate periods of poor manufacturing performance – Ability to respond to and accommodate periods of poor supplier performance – Ability to respond to and accommodate periods of poor delivery performance – Ability to respond to and accommodate new products, new markets, or new competitors
2	Gunasekaran et al. (2001)	Strategy	Total supply chain cycle time – Total cash flow time – Customer query time – Level of customer perceived value of product – Net profit versus productivity ratio Rate of return on investment – Range of product and services – Variations against budget – Order lead time – Flexibility of service systems to meet particular customer needs – Buyer–supplier partnership level – Supplier lead time against industry norm – Level of supplier's defect-free deliveries – Delivery lead time – Delivery performance

TABLE 3.4 *(Continued)*

Sr. No.	Authors	Measurement perspective	Performance measurement criteria
		Tactical	Accuracy of forecasting techniques – Product development cycle time – Order entry methods – Effectiveness of delivery invoice methods – Purchase order cycle time – Planned process cycle time – Effectiveness of master production schedule – Supplier assistance in solving technical problems – Supplier's ability to respond to quality problems – Supplier's cost-saving initiatives – Supplier's booking in procedures – Delivery reliability – Responsiveness to urgent deliveries – Effectiveness of distribution planning schedule
		Operational	Cost per operation hour Information carrying cost – Capacity utilization – Total inventory as: incoming stock level, work-in-progress, scrap level, finished goods in transit – Supplier rejection rate – Quality of delivery documentation – Efficiency of purchase order cycle time – Frequency of delivery – Driver reliability for performance – Quality of delivered goods – Achievement of defect-free deliveries

TABLE 3.4 *(Continued)*

Sr. No.	Authors	Measurement perspective	Performance measurement criteria
3	Chan and Qi (2003)	Supplying (delivery)	Delivery cost – Delivery reliability: timeliness, error-free – Delivery flexibility; frequency, amount
		Inbound logistics (transportation)	Transport cost – Transport productivity – Transport flexibility – Facility utilization
		Core manufacturing (international manufacturing operations)	Product quality – Operation cost – Efficiency – Flexibility – Productivity
		Outbound logistics (warehousing)	Warehouse costs – Inventory flow rate – Inventory accuracy – Stock capacity – Facility utilization
		Marketing and sales (customer order processing and delivery)	Response time – Order fill rate – Order flexibility; frequency, amount – Delivery reliability: timeliness, error-free

TABLE 3.4 *(Continued)*

Sr. No.	Authors	Measurement perspective	Performance measurement criteria
4	Otto and Kotza (2003)	System dynamics	Capacity utilization – Cumulative inventory level – Stock-outs – Time lags – Time to adapt – Phantom ordering
		Operations research or information technology	Logistics costs per unit – Service level – Time to deliver
		Logistics	– Integration – Lead times – Order cycle time – Inventory level – Flexibility
		Marketing	Customer satisfaction – Distribution costs per unit – Market share or channel costs
		Organization	– Transaction costs – Time to network – Flexibility – Density of relationships
		Strategy	Time to network – Time to market – ROI of focal organization

TABLE 3.4 *(Continued)*

Sr. No.	Authors	Measurement perspective	Performance measurement criteria
5	Gunasekaran et al. (2004)	Strategy - Plan	– Level of customer-perceived value of product – Variances against budget – Order lead time – Information processing cost – Net profit versus productivity ratio – Total cycle time – Total cash flow time – Product development cycle time
		Source	NIL
		Make/assembly	– Range of products and services
		Deliver	– Flexibility of service system to meet customer needs – Effectiveness of enterprise distribution planning schedule
		Tactical plan	Customer query time – Product development cycle time – Accuracy of forecasting techniques – Planning process cycle time – Order entry methods – Human resource productivity
		Source	Supplier delivery performance – Supplier lead time against industry norm Supplier pricing against market – Efficiency of purchase order cycle time – Efficiency of cash flow method – Supplier booking in procedures

TABLE 3.4 *(Continued)*

Sr. No.	Authors	Measurement perspective	Performance measurement criteria
		Make/assembly	Percentage of defects – Cost per operation hour – Capacity utilization – Utilization of economic order quantity
		Deliver	– Flexibility of service system to meet customer's needs – Effectiveness of enterprise distribution planning schedule – Effectiveness of delivery invoice methods – Percentage of finished goods in transit – Delivery reliability performance
		Operational - Plan	Order entry methods – Human resource productivity
		- Source	– Efficiency of purchase order cycle time – Supplier pricing against market
		- Make/assembly	– Percentage of defects – Cost per operation hour – Human resource productivity index
		- Deliver	Quality of delivered goods – On-time delivery of goods – Effectiveness of delivery invoice methods – Number of faultless delivery notes invoiced – Percentage of urgent deliveries – Information richness in carrying out delivery – Delivery reliability performance
6	Huang et al. (2005)	Delivery reliability	Delivery performance – Fill rates – Perfect order fulfillment

TABLE 3.4 *(Continued)*

Sr. No.	Authors	Measurement perspective	Performance measurement criteria
		Responsiveness	Order fulfillment lead times
		Flexibility	Supply chain response time
		Cost	Cost of goods sold – Total supply chain management cost – Value-added employee productivity – Warranty or return processing costs
		Assets	Cash-to-cash cycle time – Inventory days of supply – Asset turns
7	Aramyan et al. (2007)	Efficiency	– Production costs – Transaction costs – Profit – Return on investment – Inventory
		Flexibility	– Customer satisfaction – Volume flexibility – Delivery flexibility – Backorders – Lost sales
		Responsiveness	– Fill rate – Product lateness – Customer response time – Lead time – Customer complaints – Shipping errors

TABLE 3.4 *(Continued)*

Sr. No.	Authors	Measurement perspective	Performance measurement criteria
		Food quality - Product quality	– Sensory properties and shelf-life: appearance, taste, shelf life – Product safety and health: salubrity, product safety – Product reliability and convenience: product reliability, convenience
		- Process quality	– Production system characteristics: traceability, storage and transportation conditions, working conditions – Environmental aspects: energy use, water use, pesticide use, recycling or reuse – Marketing: promotions, client service, display in stores
8	Berrah and Cliville (2007)	Same as Gunasekaran et al. (2004)	Same as Gunasekaran et al. (2004)

TABLE 3.4 *(Continued)*

Sr. No.	Authors	Measurement perspective	Performance measurement criteria
9	Bhagwat and Sharma (2007a)	Finance perspective	– Net profit versus productivity ratio – Rate of return on investment – Variations against budget – Buyer–supplier partnership level – Delivery performance – Supplier's cost-saving initiatives – Delivery reliability – Cost per operation hour – Information carrying cost – Supplier's rejection rate
		Customer perspective	– Customer query time – Level of customer-perceived value of product – Range of products and services – Order lead time – Flexibility of service systems to meet particular customer needs – Buyer–supplier partnership level – Delivery lead time – Delivery performance – Effectiveness of delivery invoice methods – Delivery reliability – Responsiveness to urgent deliveries – Effectiveness of distribution planning schedule – Information carrying cost – Quality of delivery documentation – Driver reliability for performance – Quality of delivered goods – Achievement of defect-free deliveries

TABLE 3.4 *(Continued)*

Sr. No.	Authors	Measurement perspective	Performance measurement criteria
		Internal business process perspective	– Total supply chain cycle time – Total cash flow time – Flexibility of service systems to meet particular customer needs – Supplier lead time against industry norms – Level of supplier's defect-free deliveries – Accuracy of forecasting techniques – Product development cycle time – Purchase order cycle time – Planned process cycle time – Effectiveness of master production schedule – Capacity utilization – Total inventory cost: incoming stock level, work-in-progress, scrap value, finished goods in transit – Supplier's rejection rate – Efficiency of purchase order cycle time – Frequency of delivery
		Innovation and learning perspective	– Supplier's assistance in solving technical problems – Supplier's ability to respond to quality problems – Supplier's cost-saving initiatives – Supplier's booking in procedures – Capacity utilization – Order entry methods – Accuracy of forecasting techniques – Product development cycle time – Flexibility of service systems to meet particular customer needs – Buyer–supplier partnership level – Range of products and services – Level of customer-perceived value of product

TABLE 3.4 *(Continued)*

Sr. No.	Authors	Measurement perspective	Performance measurement criteria
10	Jammernegg and Reiner (2007)	Customer satisfaction improvement	– Improve delivery efficiency toward customer demands – Improve delivery efficiency toward promises – Decrease lead time for completing orders – Improve service quality – Improve response to customer demands
		Business process and information integration	– Improve purchasing process – Improve data validity – Strengthen information sharing – Increase stock turnover – Increase supply chain response – Improve efficiency for order processing – Shorten lead time for raw material providing – Decrease prediction errors – Increase flexibility of production schedule – Lower material shortage – Decrease payment time
		Organization innovation capability	– Production innovation – Cooperative design capability increasing – Innovation in production process – Diversification in products – Market entry lead time decreasing – Customization capability increasing

TABLE 3.4 *(Continued)*

Sr. No.	Authors	Measurement perspective	Performance measurement criteria
		Enterprise profit increasing	– Increase marginal profits of supply chain – Increase quality of material supply – Lower stock costs – Lower purchase costs – Lower operation costs – Increase cash turnover – Increase sales volume – Increase asset rewards ratio
		Human resource management	– Increase working efficiency – Increase employee's capability in learning new knowledge – Increase employee's adaptation for changes – Increase positive interaction between employee and the environment – Increase employee capability in independently completing a task
11	Hwang et al. (2008)	Reliability	– Perfect order fulfillment
		Responsiveness	– Order fulfillment cycle time
		Flexibility	– Upside supply chain flexibility – Upside supply chain adaptability – Downside supply chain adaptability
		Cost	– Supply chain management cost – Cost of goods sold
		Asset management	– Cash-to-cash cycle time – Return on supply chain fixed assets

TABLE 3.4 *(Continued)*

Sr. No.	Authors	Measurement perspective	Performance measurement criteria
12	Robb et al. (2008)	Operations dimension performance	– Delivery dependability – Product reliability – After-sale service – Consistent quality – Product durability – Low production cost – Production time – New products – Delivery time – New product development time – Product mix flexibility – Volume flexibility – Modification flexibility
		Human resources factors	– Improve worker safety – Motivate workers – Increase supervisor training – Provide more worker training – Give workers a broader range of tasks – Heavily involved in product design decisions from production personnel – The worker's skills at doing their own jobs – Level of training given to workers – Training workers to manage different stages of the production process – Consulting with workers in deciding the production schedule – Having no role for workers in improving the manufacturing process – Give workers more planning responsibility
13	Theeranuphattana and Tang (2008)	Reliability	– Perfect order fulfillment

TABLE 3.4 *(Continued)*

Sr. No.	Authors	Measurement perspective	Performance measurement criteria
		Responsiveness	– Order fulfillment cycle time
		Flexibility	– Upside supply chain flexibility – Upside supply chain adaptability – Downside supply chain adaptability
		Cost	– Supply chain management cost – Cost of goods sold
		Asset management	– Cash-to-cash cycle time – Return on supply chain fixed assets – Return on working capital
14	Zhu et al. (2008)	Environmental performance	– Reduction of air emission – Reduction of waste water – Reduction of solid waste – Decrease of consumption for hazardous, harmful, or toxic materials – Decrease of frequency for environmental accidents – Improvement of an enterprise's environmental situation
		Eco performance	– Decrease cost for materials purchasing – Decrease cost for energy consumption – Decrease fee for waste treatment – Decrease fee for waste discharge – Decrease fine for environmental accidents
		Operational performance	– Increase delivered on time – Decrease inventory levels – Decrease scrap rate – Promote products' quality – Increase product line – Improve capacity utilization

TABLE 3.4 *(Continued)*

Sr. No.	Authors	Measurement perspective	Performance measurement criteria
15	Cai et al. (2009)	Resource	– Total supply chain management costs – Distribution costs – Inventory costs – Manufacturing costs – Total turnover costs – Information management costs – Value-added employee productivity – Warranty costs – Return on investment
		Output	– Sales (or profit) – Rates of stock-outs (lose sales) – Fill rate: target fill rate achievement, average item fill rate – Order fulfillment lead time – Percent of on-time deliveries – Perfect of order fulfillment – Customer satisfaction – Rates of customer complaints – Planned process cycle time – Cash-to-cash cycle time

TABLE 3.4 *(Continued)*

Sr. No.	Authors	Measurement perspective	Performance measurement criteria
		Flexibility	– Supply chain responsiveness – Manufacturing flexibility – Procurement flexibility – Logistics flexibility – Delivery flexibility – New products flexibility – Information systems flexibility
		Innovativeness	– Rates of sales in new products – Number of new products launched – Process improvement
16	Cai et al. (2009)	Innovativeness	– Supply chain stability
		Information	– Information accuracy – Information timeliness – Information availability – Information sharing
17	Chae (2009)	Sales and marketing	– Forecast versus order – Forecast volatility – Inventory days of supply at sales subsidiaries
		Production	– On-time departure from manufacturing subsidiaries – Production plan versus result – Inventory days of raw material supply – Inventory days of (finished goods) supply at manufacturing subsidiaries – On-time arrival to sales subsidiaries (or distribution centers) from – manufacturing subsidiaries

TABLE 3.4 *(Continued)*

Sr. No.	Authors	Measurement perspective	Performance measurement criteria
		Purchasing	– Supplier fill rate – Automatic purchasing order (PO) rate
		Operation strategy	– Forecast accuracy – Planning cycle – Inventory days of supply – Cash-to-cash cycle
18	Chia et al. (2009)	Financial perspective	– Return on investment – Gross revenue – Profit before tax – Cost reduction
		Customer perspective	– Market share – Number of customers retained – Customer satisfaction
		Business processes perspective	– Quality of services – New services implemented per year – On-time delivery – Waste reduction
		Learning and growth perspective	– Employee satisfaction – Employee turnover per year – Number of suggestions implemented per employee yearly – Money invested in employee training yearly
19	Rodriguez et al. (2009)	Financial perspective	– Net benefit (sales) – Turnover increment – Costs of structure – Financial indebtedness – Final stock of cloths per campaign

TABLE 3.4 *(Continued)*

Sr. No.	Authors	Measurement perspective	Performance measurement criteria
		Customer perspective	– Customer average order – Number of customers that ask for repetition divide by number of total customers – Number of repeating requests per model and customer divide by number of customers that ask for repetition – Percentage of annulled models – Efficacy of the model – Global efficacy of each collection per intermediary – Average order per intermediary – Number of new customers captured – Number of orders per week (sales speed) – Increment of customers – Number of captured customers that are already customers of other enterprise of the organization divide by number of new customers – Number of nonaccomplished orders divided by total number of orders – Number of customers that make a claim divided by total number of customers – Number of claims per model and customer divided by total number of customers that make a claim – Increment of incidences occurred between the last 2 years – Average reaction time to one customer's claim

TABLE 3.4 *(Continued)*

Sr. No.	Authors	Measurement perspective	Performance measurement criteria
		Internal perspective	– Stock variability – Percentage of return item per model – Percentage of customer acceptation – Total number of exited models divided by total number of produced models – Number of clothes with deviations with respect to the standard scaled – Customer accomplishment degree – Deviation between the order date and the reception date – Total number of innovative clothes divided by total number of produced clothes – Number of innovative successful clothes divided by total number of innovative clothes – Number of reprocessed clothes divided by total number of clothes – Number of customers buying a determined style divided by total number of customers – Number of annulled catalogues divided by total number of catalogues
		Learning and growth perspective	– Number of critical processes totally documented divided by total number of critical processes – Number of critical processes controlled by the follow-up tool divided by total number of critical processes – Number of yearly training hours divided by number of yearly work hours – Hours needed of real training (detected by the enterprise) divided by hours needed of theoretical training (that employees think they should have) – Number of established inter-areas work or communication protocols divided by number of essential processes – Establishment of inter-areas work or communication protocols (Yes/No) – Number of inter-areas personnel interchanges divided by number of areas susceptible of interchange within the enterprise – Development of an inceptive plan (Yes/No) – Number of employees that know exactly the objective to be reached divided by total number of employees

TABLE 3.4 *(Continued)*

Sr. No.	Authors	Measurement perspective	Performance measurement criteria
20	Thakkar et al. (2009)	Customer service	– Service quality – Product quality – Timely delivery – Responsiveness – Order fill rate – After sales service – Perfect order fulfillment
		Finance and marketing	– Profit margins – Pretax return on assets – After tax return on investment – Return on investment – Return on assets – Total supply chain cost – Growth in market share – Return on capital employed – Improved cash flow – Warranty or returns processing cost
		Internal business	– Inventory turnover ratio – Throughput time – Percentage scrap – Accuracy of documentation – On-time delivery – Unit cost reduction

TABLE 3.4 *(Continued)*

Sr. No.	Authors	Measurement perspective	Performance measurement criteria
		Innovation and learning	– Flexible work force – Product innovation – Process innovation – Information sharing across supply chain – Training to managers and workers – Vendor development initiatives – Design modification based on customer requirements
21	Bigliardi and Bottani (2010)	Financial perspective	– Information carrying cost – Supplier's cost-saving activities – Variations against budget – Cost per operation hour – Return on investment
		Customer perspective	– Customer query time – Order lead time – Distribution lead time – Distribution performance – Delivery reliability – Effectiveness of distribution planning schedule – Quality of delivery goods – Customer-perceived value of product – Flexibility of service system to meet particular customer needs – Responsiveness to urgent delivery

TABLE 3.4 *(Continued)*

Sr. No.	Authors	Measurement perspective	Performance measurement criteria
		Internal process perspective	– Accuracy of forecasting techniques – Purchase order cycle time – Planned process cycle time – Effectiveness of master production schedule – Supplier's rejection rate – Total inventory cost – Frequency of delivery
		Learning and growth perspective	– Supplier assistance in solving technical problems – Supplier's ability to respond to quality problems – Buyer–supplier collaboration in problem solving – Order entry method – Level of information sharing
22	Flynn et al. (2010)	Customer integration	– Level of linkage with customer through information networks – Level of computerization for customer's ordering – Level of sharing of market information from customer – Establishment of quick ordering systems with customer – Level of communication with customer – Follow-up with customer for feedback – Frequency of period contacts with customer – Sharing point of sales (POS) information of major customer with company – Sharing demand forecast of major customer with company – Sharing available inventory of company with major customer – Sharing production plan of company with major customer

TABLE 3.4 *(Continued)*

Sr. No.	Authors	Measurement perspective	Performance measurement criteria
		Supplier integration	– Level of information exchange with supplier through information networks – Establishment of quick ordering systems with supplier – Level of strategic partnership with supplier – Stable procurement through network with supplier – Participation level of supplier in the process of procurement and production – Participation level of supplier in the design stage – Sharing production schedule of major supplier with company – Sharing production capacity of major supplier with company – Sharing available inventory of major supplier with company – Sharing production plan of company with major supplier – Sharing demand forecasts of company with major supplier – Sharing inventory levels of company with major supplier – Helping major supplier to improve process to better meet company's needs
		Internal integration	– Data integration among internal functions – Enterprise application integration among internal functions – Integrative inventory management – Real-time searching of the level of inventory – Real-time searching of logistics-related operating data – Utilization of periodic interdepartmental meetings among internal functions – Use of cross functional teams in process improvement – Use of cross functional teams in new product development – Real-time integration and connection among all internal functions from raw material management through production, shipping, and sales

TABLE 3.4 *(Continued)*

Sr. No.	Authors	Measurement perspective	Performance measurement criteria
		Operational performance	– Quickly modify products to meet major customer's requirements – Quickly introduce new products into the market – Quickly respond to changes in market demand – Outstanding on-time delivery record to major customer – Lead time for fulfilling customers' orders – Provide a high level of customer service to major customer
		Business performance	– Growth in sales – Return on sales – Growth in return on sales – Growth in profit – Growth in market share – Return on investment (ROI) – Growth in ROI
23	Lin et al. (2010)	Innovation	– Flexibility of delivery systems to meet the customer needs – Strengthen the supplier partnerships – Cost competitiveness – Shorter order cycles – Flexible customer response

Bhagwat et al. (2007a) classifying under four perspectives of balanced scorecard (financial, customer, internal business process, and innovation and learning). Also, the comprehensive review of literature highlights the gaps in developing the supply chain performance measurement system for textile industry. The focus of this study is on developing the performance evaluation framework for textile supply chain in developing countries, which is at a very nascent stage. The consideration of these aspects was because the facets of organization differ with respect to economical stability of the nation.

3.6 SUPPLY CHAIN PERFORMANCE MEASUREMENT IN TEXTILE INDUSTRY

As a matter of fact, in developing countries like India, textile and clothing industries set up one of the pillars of backbone of economy. Due to world trade organization's decision on the abolition of quotas in 2005, textile industries are facing intense competition from global players. Following trend among industries, their supply chains need to cope with global competition for achieving efficient supply chain. To achieve effectiveness and efficiency, textile industries have to employ new strategies to sustain. The most prominent philosophy is such case is strategic management and planning. Strategic management is concept of decision-making suiting to the satisfaction of all stakeholders of an industry concerning developing, implementing, and controlling the performance of each player in the supply chain. To effectively design strategies for the industry, one must look for the SWOT analysis of the industry, based on which the performance attributes would be decided. The SWOT analysis for the textile industry is as follows:

Strength:

1. Autonomous and self-dependent industry.
2. Ample availability of raw material that facilitates in reduction of costs and lead time.
3. Low-cost manpower leads to economic competitiveness.
4. It has great advantage of spinning sector which is available in all processes of operation in supply chain.
5. It is having manufacturing flexibility which helps improved productivity.

Weakness:

1. Highly scattered industry, leading to improper execution of cyclic process of its supply chain.
2. Lower segmental financial productivity.
3. It has technological obsolescence resulting in affecting the production efficiency and also other functions of its supply chain.
4. Infrastructural bottlenecking with respect to their cyclic processes.
5. Inefficient transaction time in distribution ports and distribution channels.

Opportunities:

1. Big potential domestic and international markets.
2. Product development using high-end software packages to satisfy diversified global needs.
3. Market shift toward branded readymade clothing.
4. Huge opportunity due to emerging retail market and malls.
5. Incorporate use of information and communication technology to enhance information sharing.

Threats:

1. Sustain in global competition among industry's supply chain.
2. Continual improvement program for quality enhancement is needed.
3. Designing workplace effective for the worker involved, that is, skill enhancement, training, etc.
4. Demand versus supply balance.
5. Price versus quality balance.

The important issue after the development of the strategy is implementation. To sustain in competitive global supply chain, industries must design strategies, develop strategic objective or goals in line with strategies, and allocate specific and suitable performance attributes to these objectives. In literature review it is found that studies are being conducted to evaluate the sourcing strategies in textile and apparel industry, where cost was a prime factor to evaluate the vendor performance (Sauls, 2007). But considering financial aspects such as cost would not be sufficient to provide effectiveness and efficiency to textile industries. Following one study is conducted where service to customer was found to be the important parameters in the textile industry.

Hence, to achieve best of the quality and best practices among industry or supply chain and to convert the weaknesses into strength, standard and comprehensive performance criteria must be selected for the textile supply chain. Kaplan and Nortan (1996) have provided balanced scorecard (BSC), a comprehensive set of performance perspectives which elaborates the performance of supply chain in a more strategic manner and hence known as a strategic performance management system.

From a literature review and SWOT analysis, BSC perspectives (financial, customer, internal business perspective, and innovation and learning perspective) are selected as performance criteria with the help of which the performance of the textile supply chain is evaluated.

Business strategy must have synergy with supply chain strategy (Sanders, 2012). To make it possible, it is proposed to align SCM to balanced scorecard perspective. Supplier selection and evaluation is a prominent area of research, whereas topics like SCPM, supply chain quality management, supply chain risk management, supply chain flexibility, etc., were less focused. There are few attempts made to implement SCPM system in textile industries. Below mentioned in Table 3.5 are two studies critically relevant to textile industries, which considered involvement of various performance measures. With a view to overcoming the limitation in these studies, our study proposes to combine BSC perspectives, various performance measures, and supply chain cyclic processes to develop an integrated framework for the textile industry.

3.7 SUMMARY

The literature review has laid down the blueprint for developing the key areas to understand views concerning supply chain performance models and their performance measurement criteria. It addressed the issue regarding textile supply chain management in India. As mentioned in the literature, the modern performance measurement frameworks evaluate performance considering many aspects. Model evaluates performance across supply chain activities (e.g., SCOR model), also among decision-making levels (strategic, tactical, and operational). To summarize, the present study categorizes the performance measurement frameworks for supply chain into three major models: outcome based (balanced scorecard), activity based (supply chain operations reference, SCOR), and last hierarchical (decision-making levels) models. Every framework has advantages and disadvantages as listed earlier in Table 3.5. Studies with

TABLE 3.5 Featured Details of Studies Conducted for the Implementation of SCPMS in Textile Industries.

Geographical location	Focused part of textile supply chain	Factors considered for framework	Methodology adopted	Limitation	Future scope
Srilanka	Downstream supply chain players (garment manufacturers).	Supplier, internal operation, and supply chain.	Two phases (field research) and (exploratory study consists of Internet survey and phone interviews) study is conducted. Analysis was done using Chi-square test.	Upstream players are not considered. Customer aspect not touched. Comprehensive framework is lacking. Fabric formation step is omitted.	Structured performance measures may be identified for large and small companies. Study can be conducted aligning business strategy to supply chain processes.
South India	Considered apparel manufacturers only.	Four dimensions - Quality, flexibility, dependability, and innovation.	Hypotheses were framed Data collected and analyzed using correlation, MANOVA, and multiple regression.	Earlier supply chain partners not focused customer orientation is lacking. Structured framework is lacking. Responses sought from production managers only.	Study may involve various cyclic processes. Value creation perspective can be considered for getting holistic view. Data can be collected from managers of other divisions also.

respect to textile supply chain performance lack in developing a comprehensive framework which facilitated the measurement of supply chain performance as a whole. Hence, the proposed studies wished to bridge the gap and intended to provide a more holistic approach to performance measurement framework for supply chain management in textile industries in India. The approach would be attained by aggregating the supply chain cyclic process, balanced scorecard perspective and selection of performance measurement factor under each perspective would be classified as different decision-making level.

The suggested supply chain performance measurement framework consisting of performance measurement criteria categorized under four BSC perspectives will be evaluated on supply chain cyclic processes (procurement–production–distribution) which integrates supplier, manufacturer, and immediate customer. The proposed model helps organizations to design a right supply chain strategy by analyzing the contribution of supply chain cyclic processes toward overall supply chain performance of textile industry. The research plan of work is elaborated in the next chapter. It overviews the viewpoint and approaches which guide the research structure, methodology, and techniques.

KEYWORDS

- **supply chain performance measurement**
- **textile industry**
- **comparative analysis**
- **performance measurement criteria**
- **key performance**

CHAPTER 4

SUPPLY CHAIN PERFORMANCE MEASUREMENT FRAMEWORK: RESEARCH CASE STUDY IN TEXTILE INDUSTRIES

ABSTRACT

This chapter provides an overview of the stages involved in the development of a supply chain performance measurement (SCPM) framework for the textile industry. It summarizes the methodologies employed in each stage, the result of the first stage, the purpose of the second level of the Delphi method, the conclusion of the third level of the Delphi method, and the identification of case companies for implementation. The development of an SCPM framework entails a structured process to create a comprehensive system for evaluating and monitoring SC performance in the textile industry. The first stage involves the implementation of a methodology that guides the framework's development. In the methodology's first stage, various techniques such as literature review, expert opinions, and industry analysis are employed to gather relevant information and identify key performance indicators (KPIs) specific to the textile industry. This ensures that the framework is tailored to the industry's unique requirements. The result of the first stage is the development of a SCPM system (SCPMS) specifically designed for textile industries. This system encompasses a set of KPIs and measurement tools that enable companies to assess and track their SC performance effectively. The second stage

Supply Chain Performance Measurement in Textile Enterprises. Pranav C. Charkha, PhD, Santosh B. Jaju, PhD & Prasenjit Chatterjee, PhD (Authors.)

utilizes the Delphi method, a structured communication process involving a panel of experts, to refine and validate the SCPMS. The purpose of the second level of the Delphi method is to obtain expert consensus on the relevance and effectiveness of the SCPMS components, ensuring its robustness and applicability in real-world scenarios. The third level of the Delphi method concludes the expert consensus process, ensuring that all relevant perspectives are considered and any discrepancies are addressed. This level provides a comprehensive and refined SCPMS that incorporates the collective expertise and insights of industry professionals.

4.1 SCPM FRAMEWORK DEVELOPMENT

4.1.1 INTRODUCTION

As explained in earlier chapter, the ultimate aim of the research study is to provide a set of performance measures for textile supply chain performance measurement. To attain the desired target, two phases are proposed and presented step-by-step. The aim of first stage (modified Delphi technique) is to develop supply chain performance measurement framework for textile industry. The aim of second stage is to evaluate the set of performance measures in the framework by validating them for their applicability to four different textile industries. The stepwise plan of work is shown in Figure 4.1.

4.2 METHODOLOGY IN STAGE ONE

4.2.1 LITERATURE REVIEW

In order to attain the research aim and objective, an extensive review of the literature is conducted on the issues concerning supply chain management and performance measurement system with special consideration on the textile industry in India. Our two main areas of search included: (1) supply chain performance measurement models and (2) performance measurement factor/criteria and attributes also termed as performance metrics or key performance indicators (KPIs). More than 100 academic research papers along with government reports of progressive 5 year plans were reviewed. The literature reviewed was from 1995 to 2010 using database from peer-reviewed articles of international journals' database of Inderscience, Emerald, ScienceDirect, and Elsevier; journals related to textile such as Fiber to fashion, Textile journal of India, etc., were also searched. These researches were studied and reviewed to categorize and recognize what work has been done in order to identify the gaps. This pool of knowledge as explained in earlier chapter was appropriate and gave us ancestral justification for theory building or work extension in the area of supply chain performance measurement in textile industries, preferably in India.

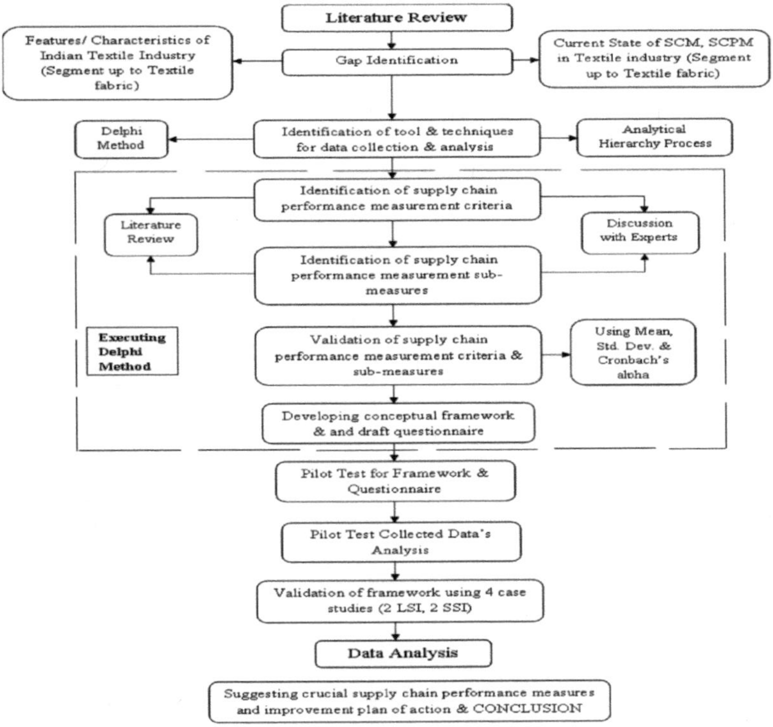

FIGURE 4.1 Flow chart of research work.

4.2.2 DELPHI METHOD

The Delphi is a well-known structured, qualitative, and featured method using indirect interaction in usages today, developed by Olaf Helmer and Norman Dalkey in 1953 in RAND Corporation to address some issues of the military. It became more popular with the application of this technique to major forecasting and corporate planning (Helmer, 1983).

The Delphi technique has several advantages as reported in literature reviews such as anonymity, written responses, controlled feedback, and statistical group responses. One important feature is that respondent need not to interact face-to-face, interviews may remain in absentia, wherein sufficient amount of time is allotted to respondent for thinking and submitting feedback. Sufficient flexibility is allotted to respondents and issue of allegiance is avoided. It is fast, inexpensive, easy to implement, and dynamic in nature in the matters wherein importance of expert opinion is pursued.

The Delphi technique consists of multiple sequential rounds of survey questionnaire. In the conventional Delphi method, during the first round, respondents are asked to reply a broad area question related to field, whereas each progressive round would give more gleaned responses. Reaching to a common consent leads to the termination of the process. As per the analysis of Linstone and Turoff (1975), generally, three rounds are adequate to reach a common consensus in the responses. Rounds extended thereafter, not lead to any significant changes and thereof, prove unwanted to respondents.

The detailed steps of the Delphi technique, as suggested by Pawlowski (2004), are as follows:

- Expert panel selection—Expert in relevant field of study must be selected for executing the discussion. The panel should consist of 10–30 numbers of experts. Due to time-consuming task, size of panel with 10–15 is recommended.
- First round of discussion, where questions pertaining to knowledge of undertaken study are very broad and open-ended, and no structured or restricted format is used by the researchers.
- Responses were analyzed and put into specific categories.
- Second round is executed to develop questionnaire and responses, analyzed and ranked from experts for feedback.
- Third round questionnaire is then developed, where panel was asked to review the median scores from earlier round and analyzed the responses again for major changes. During this, if any experts differ from group median, they were requested to provide specific remarks for the same.

- Finally, median scores were calculated to determine whether consensus was reached.
- At last, report was generated containing mean and median scores, and Cronbach's alpha to find the degree of agreement and consensus.

4.2.3 MODIFIED-DELPHI TECHNIQUE

To undertake the application of the Delphi technique, slight modification is done while executing the three phases of the method. The change is carried out in stage first. Panel was formed with 15 persons comprising experts from textile industry, government, and local association; experts from textile engineering institutes and practitioners working in the field of industrial engineering and operation management, preferably from the central India. All were identified by contacting them personally and their consent was sought. During the execution of first stage, two members of the panel were finding difficult to spare time due to their preoccupied event and responsibility. Hence, at last, 13 members were finalized and the panel was finalized.

As stated earlier, modification is done at very initial stage. The modification was replacement of the conventional pattern of questionnaire by one-to-one in-depth interviews to establish communication with the experts in the panel. The replacement of questionnaire survey of experts with one-to-one in-depth interview is because, from an extensive literature review, we came to know that research in supply chain performance measurement with respect to the textile industry, especially in India, is at a very nascent stage. The studies for supply chain performance evaluation in textile industry are at very initial stage of its theoretical and practical use in developing countries like India. This also leads to the conclusion that even the experts of the textile industry in India are aware of this concept to a little extent, as the textile supply chain is very complex and consists of various partner firms such as ginning, spinning, weaving, processing, and finishing. The development philosophy of the supply chain performance measurement system (SCPMS) and the use of SMART metrics for supply chain performance assessment are assumed to be difficult to deliver through questionnaire surveys. Hence, it is not advisable to initiate the discussion regarding SCM and SCPMS through questionnaire and for this reason, face-to-face interview is proposed and proved to be more efficient and accurate, as can be controlled and closely monitored by the interviewer. The explanation regarding some intricate terminologies can be easily provided to interviewees. The only constraint in face-to-face interview is to arrange the schedule of interviews with experts during their busy schedule. It takes a long time to convey the issue or matter pertaining to SCM/SCPMS in the textile industry. Each interview was conducted for about 1 h. A detailed plan was prepared for smooth conduction of the interview which includes open-ended questions based on the literature review. At an initial stage, expert panel was asked to submit its opinion on generalized issues like SCM and SCPM in the Indian textile industry.

The process of interview was executed from March 2013 to June 2013. The responses were noted down, analyzed, and sent for survey of second round to 13 experts during end of July 2013 in person using hard copy. Experts took around 1 month to reply, which involved reminder telephone calls made to request all of them to complete the questionnaire.

After collecting all the questionnaires, responses were summarized and final supply chain performance measurement system employed in this study was developed. Finally, results were sent via mail to experts for consenting confirmation in the third round.

Detailed list of experts for the Delphi technique questionnaire is presented in Appendix A and B, respectively.

4.3 METHODOLOGY IN STAGE TWO

4.3.1 IDENTIFICATION OF CASE TEXTILE COMPANIES FOR APPLICATION OF FRAMEWORK

The study is focused on identifying the practices adopted in the textile industry for "*supply chain management*" with special consideration of its "*performance measurement.*" The study identified four industries, contributing the diversity of two sizes of industry. Out of four identified industries, two are large-scale units, whereas other two are small and medium enterprises. All four case companies are located in the central region of India in Maharashtra.

The selected industries are well and sufficient to give insights into textile supply chain management. The two large-scale units are producing best quality ready to stitch fabric having own retail outlets and also providing products to almost all the major apparel manufacturers, whereas the other two small-scale units are producing the unfinished fabric and yarn to supply to the sophisticated garment manufacturing and weaving units, respectively.

The two selected large-scale units in the sample are jointly operating businesses with their counterpart of foreign origin and hence experience the impact of similar magnitude of select issues such as rapid development of technologies and requirement for growth in the competitive volatile global markets.

4.3.2 SELECTION OF INTERVIEWEES (INDUSTRY PERSONNEL)

Depending on the framework developed, this work targeted to identify crucial performance measurement subcriteria associated with the four performance perspectives (criteria) of an organization (*financial, customer, internal business, and innovation and learning*). Thus, the target interviewees for the study were selected based on the years of experience, knowledge of supply chain management, and performance measurement system. The respondents were categorized into six categories: respondents from procurement department representing relationship with suppliers, respondents from the production department giving the status of internal operation of an organization, respondents from distribution department representing the relationship with customers (wholesaler or retailer), respondents from the quality control department signifying the efforts done to impart various quality improvement initiatives/techniques, respondents from human resource department (HRD) giving an insight into what is best and training/learning adopted for employees to make them aware about the evaluation of various types of methodologies in the context of supply chain management, and respondents from commercial department handling issues like cost accounting, forecasting, etc.

Generally, qualitative samples are smaller in size. Ritchie et al. (2003a) quoted that there is nothing mandatory to ensure that the selected sample is of sufficient quantity to provide exact results, or to determine statistically important biased variables. Instead, the type of information that qualitative studies yield is rich in details. Yet, in this study, the selected sample is of sufficient size so as to provide the detailed insights of the textile supply chain performance measurement practices in both types of organizations.

Therefore, the number of interviewees with different types of actors could be 80 as shown in Table 4.1.

TABLE 4.1 Interviewers Identified from Four Case Companies.

Interviewing group	Actor	Number of interviewers
Top management	Managing Director General Manager/Factory Manager	02
Commercial department	Manager Asst. Manager Executive/Supervisor	03
Procurement department	Manager Asst. Manager Executive/Supervisor	03
Production department	Manager Asst. Manager Executive/Supervisor	03
Distribution department	Manager Asst. Manager Executive/Supervisor	03
Quality control department	Manager Asst. Manager Executive/Supervisor	03
Human resource department	Manager Asst. Manager Executive/Supervisor	03

4.3.3 WHY ANALYTIC HIERARCHY PROCESS USED IN STAGE TWO

The concept of supply chain performance measurement possesses much ambiguity and vagueness when it is concerned with textile industry, especially in developing country like India. Attempts have been made to evaluate the performance of supply chain using hypothesis test, cost accountability, etc. But, they lack in providing the concrete group of performance measure/metrics and performance measurement criteria. As, we all know, and mentioned earlier in Chapter 2 that study pertaining to the supply chain performance measurement system in the Indian textile industry is at a very nascent stage, this confusion has hampered the competitiveness of textile supply chain. Hence, to overcome ambiguity, some issues that need to be addressed are as follows:

- ✓ **Imperceptible information**: In the recent global scenario of supply chain performance measurement system in the textile industries, quantitative and qualitative measures do exist. Qualitative terminologies acting as measures, such as "*implementation of knowledge management*" for the enhanced view of supply chain management, are not possible to be quantified accurately.
- ✓ **Insufficient information**: Unavailability of specific tool/technique to evaluate certain parameters in textile supply chain performance, for example, "*supplier's cost saving initiatives*." It is information provided with approximation and inadequacy, resultant of lack of precise tool/experts.
- ✓ **Unattainable information**: For this, the participation of supply chain function processes and supply chain partners in textile supply chain are important. Generally, exact data is rarely obtained, due to high cost of information sharing, time constraint, etc. The best example for this is estimation of "*total supply chain cycle time*." It can be done by setting up of standard work measurement setups and benchmarks by all associated divisions and partners, which enables recording in very systematic fashion. Such data can be easily used in decision-making of an organization.
- ✓ **Fractional unawareness**: It is obsceneness added to unawareness of some type of measures for their adoptability with the knowledge of half concept. Example is "*use of quality engineering/quality management techniques*."

As discussed earlier, analytic hierarchy process (AHP) is capable of handling uncertainty when coupled with Delphi feedback technique. It facilitates pairwise comparison of elements at every level. It doesn't require any large-scale assumptions; it helps in decision-making of the degree of importance of specific parameters. AHP is followed and employed in this study in the below mentioned manner.

Quantitative data, by comparing the importance of each performance measurement factor at every level in the designed framework, is collected for measuring performance of supply chain for all the case companies by using AHP. The AHP is a multicriteria decision-making method designed by Saaty (1990). It is a structured procedure to represent the attributes associated to that problem, hierarchically. Jagdev et al. (2004) quoted that as decision-making for strategy framing area is generally filled with uncertainty, it becomes important for any decision-maker/manager that the person should convey a maximum extent of belief and assurance in judgment. AHP facilitates such theory and a subsequent methodology, which provides an approach for modeling of unstructured problems.

4.3.3.1 STEPS INVOLVED IN AHP

- Problem decomposition and hierarchy construction: The top level of the hierarchy is the overall objective. The next level is the criteria. Below this level are the subcriteria.
- Determine alternatives: The decision alternatives are constructed and added to the lowest level of the hierarchy.
- Pairwise comparison: Pairwise comparison aims at determining the relative importance of the elements in each level of the hierarchy. It starts from the second level and ends at the lowest level. The decision-maker needs to express the preference between each pair of the elements.
- Weight calculation: Mathematical normalization methods are used to calculate the priority weights for each level.
- Consistency check: A consistency ratio (CR) is calculated. If it is greater than 10%, then the decision-maker is not consistent in making the pairwise comparison. The decision-maker should review the comparison and make adjustments.
- Hierarchical synthesis: The calculated priority weights at different hierarchy levels are integrated to allow overall evaluation of the alternatives.
- Determine priority for all alternatives: The alternative with the highest overall priority weight is chosen (Huan et al., 2004).

The AHP was introduced to reveal the manner people naturally believe and think. It is a decision-making tool which allows general decision process by segregating a complex problem into a various levels of hierarchical organization of objectives, criteria, subcriteria, and alternatives. AHP's hierarchic structure reflects the usual inclination of human thinking to categorize parameters of an organization into various levels and make a group of elements in each level as per their characteristics and features, which helps decision-makers to easily understand the human point of view (Huan et al., 2004).

Abdul-Hamid (1999) proposed following advantages of AHP method:

- Hierarchical structure facilitates the subjective decision procedure which leads to accurate decisions.
- Consistency is ensured in decision judgment.
- Decomposing of problems helps in clearly understanding the problem.
- Repetitive groupwise comparative judgments are made until conformity is reached.
- AHP facilitates sensitivity analysis using sophisticated computer packages prior to final judgment to ensure the result confidentiality.

In the last two decades, AHP is extensively used for a variety of applications in the field of operations management decision formulation (Partovi et al., 1990), for estimation of risk in projects management (Dey et al., 1994), finalizing the plant layouts for facility planning (Dweiri and Meirer, 1996), implementing the quality function deployment (Bergquist and Abesysekera, 1996), new product forecasting (Calantone, 1999), making the groupings of part for machine scheduling (Ziilal and Arikan, 2000), benchmarking the standards for optimizing the parameters for project management (Dey, 2002), performance evaluation (Jagdev et al., 2004; Bhagwat and Sharma, 2007, 2008), project evaluation and selection (Dey, 2004, 2006), material and supplier selection (Dweiri and Al-oqla, 2006), operational performance measurement (Dey et al., 2006), site selection (Dey and Ramcharan, 2008), identification of performance measures (Gunasekaran, 2001, 2004, 2007; Felix and Chan, 2003), six sigma implementation (Laosirihongthong, 2005), and green supply chain performance assessment (Kurien and Quereshi, 2012).

This research study demonstrated the application of AHP as a tool to measure the performance of supply chain of the participating textile companies in India.

4.4 APPLICATION OF DELPHI METHOD

In earlier chapter of research methodology, a two-step procedure adopted for this study is elaborated. As stated earlier, after reviewing the literature and following Delphi approach, a hierarchical framework for performance evaluation of textile supply chain is built. Secondly, selection of industry for validating the application of proposed framework is done. Finally, the process of judgment comprises of application of multicriteria decision-making approach using AHP. In this study, AHP is used to execute pairwise comparison to prioritize the first-level criteria, subcriteria, and alternatives with respect to overall goal of the framework. Also, a pairwise comparison is carried out among the elements at the same level of hierarchy for understanding the importance at that level with respect to higher level hierarchy and overall supply chain performance. For getting the responses from the industry persons, questionnaire finalized after the Delphi method is used. The completely filled questionnaire is responded by 75 experts from the four case companies. A general comparison is drawn for the analysis by generalizing the result obtained for the textile industries matching the specification of the selected companies. First stage of research contributes to the development of the hierarchical framework for supply chain performance measurement in the textile industries using the Delphi method. Contribution of second stage is the importance of identified performance criteria under each of balanced scorecard (BSC) perspective. This stage gives the crucial performance measures responsible for making the textile supply chain effective and efficient.

4.5 RESULT OF STAGE ONE: DEVELOPMENT OF SUPPLY CHAIN PERFORMANCE MEASUREMENT SYSTEM (SCPMS) FOR TEXTILE INDUSTRIES

4.5.1 ANALYSIS FROM LITERATURE REVIEW

Extensive literature review has yielded two major aspects, (1) four performance perspectives as criteria on which supply chain performance for the textile industries in India can be evaluated and (2) cyclic processes of the supply chain whose contribution toward overall supply chain performance is to be estimated.

The four areas, where supply chain performance can be measured, become the first-level criteria for hierarchy; they are financial level, customer level, production operation, and training and learning. The four areas are identified from the literature reviewed pertaining to textile supply chain management and economic competitiveness of textile supply chain. These four aspects, when studied sophistically become the strategic perspectives of BSC. Hence, they were refined further to suit the standard model available for the outcome-based approach of supply chain performance measurement, that is, BSC, and refined aspects are financial perspective, customer perspective, internal business process perspective, and innovation and learning perspective.

As a matter of fact, in developing countries like India, the textile and clothing industries set up one of the pillars of the backbone of the economy. Due to world trade organization decision on abolition of quotas in 2005, the textile industries are facing intense competition from the global players. Following trend among industries, their supply chains need to cope with global competition for achieving efficient supply chain. In order to achieve effectiveness and efficiency, the textile industries have to employ new strategies to sustain. The most prominent philosophy in such case is strategic management and planning. Strategic management is a concept of decision-making suiting the satisfaction of all stakeholders of the industry concerning developing, implementing, and controlling the performance of each player in the supply chain. In order to effectively design strategies for the industry, one must look for the SWOT analysis of the industry. Based on which the performance attributes would be decided. The SWOT analysis for textile industry is as follows:

Strength:

- ✓ Autonomous and self-dependent industry.
- ✓ Ample availability of raw material that facilitates in reduction of costs and lead time.
- ✓ Low-cost manpower leads to economic competitiveness.
- ✓ It has great advantage of spinning sector which is available in all processes of operation in the supply chain.
- ✓ It is having manufacturing flexibility which helps in improved productivity.

Weakness:

- ✓ Highly scattered industry, leading to improper execution of cyclic process of its supply chain.
- ✓ Lower segmental financial productivity.
- ✓ It has technological obsolescence, resulting in affecting the production efficiency and also other function of its supply chain.

- ✓ Infrastructural bottlenecking with respect to their cyclic processes.
- ✓ Inefficient transaction time in distribution ports and distribution channels.

Opportunities:

- ✓ Big potential domestic and international market.
- ✓ Product development using high-end software packages to satisfy diversified global needs.
- ✓ Market shift toward branded readymade clothing.
- ✓ Huge opportunity due to emerging retail market and malls.
- ✓ Incorporate the use of information and communication technology to enhance information sharing.

Threats:

- ✓ Sustain in global competition among industry's supply chain.
- ✓ Continual improvement program for quality enhancement is needed.
- ✓ Designing workplace effective for the workers involved, that is, skill enhancement, training, etc.
- ✓ Demand versus supply balance.
- ✓ Price versus quality balance.

The important issue after development of a strategy is implementation. In order to sustain in competitive global supply chain, industries must design strategies, develop strategic objective or goals in line with strategies, and allocate specific and suitable performance attributes to these objectives. In a literature review, it is found that studies are being conducted to evaluate the sourcing strategies in textile and apparel industry, where cost was the prime factor to evaluate the vendor performance (Sauls, 2007). But, considering financial aspect such as cost would not be sufficient to provide efficiency to the textile industries. Following one study is conducted, where service to customer was found to be the important parameter in the textile industry.

Hence, to achieve best of the quality and best practices among industry or supply chain and in order to convert the weaknesses into strength, standard and comprehensive performance criteria must be selected for the textile supply chain. Kaplan and Nortan (1996) has provided BSC, a comprehensive set of performance perspectives, which elaborates the performance of supply chain in more strategic manner and hence known as strategic performance management system.

From literature review and SWOT analysis, BSC perspectives (financial, customer, internal business perspective, and innovation and learning perspective) are selected as performance criteria with the help of which the performance of textile supply chain is evaluated.

4.5.2 PERFORMANCE MEASURES/METRICS FOR SUPPLY CHAIN MANAGEMENT

After an extensive review of literature and stated earlier in literature review section, 35 measures were selected categorized under each of BSC perspective. The basis behind selecting the performance measures is to give the performance measurement framework a comprehensive view in terms of incorporating the decision-making levels (strategic, tactical, and operational) into it. Hence, the standard model proposed by Gunasekaran (2001, 2004) becomes the reference for selecting the performance measures. The performance measures included in this model reflect both quantitative and qualitative approach of measuring performance. The applicability of these performance measures to the textile industry would be tested in Delphi method.

The list of selected performance measures and classified under each BSC perspective is as shown in Table 4.1. Conceptual framework is then developed as shown in Figure 4.2.

TABLE 4.2 List of Performance Measures Identified and Classified Under Four BSC Perspectives.

Sr. No.	Balanced scorecard perspectives	Identified measures	Decision-making level	Nature of measures (quantitative/qualitative)
1	Financial perspective	Net profit versus productivity ratio	Strategic	Quantitative
		Rate of return on investment	Strategic	Quantitative
		Supplier's cost saving initiatives	Tactical	Qualitative
		Supplier's rejection rate	Operational	Quantitative
		Information carrying cost	Operational	Quantitative
		Cost per hour of operation	Operational	Quantitative

TABLE 4.2 *(Continued)*

Sr. No.	Balanced scorecard perspectives	Identified measures	Decision-making level	Nature of measures (quantitative/ qualitative)
		Manufacturing cost	Strategic and Operational	Quantitative
		Variation against budget	Strategic	Quantitative
		Inventory carrying cost	Operational	Quantitative
2	Customer perspective	Quality of goods delivered	Operational	Qualitative
		Effectiveness of delivery invoice methods	Tactical	Qualitative
		Delivery performance and lead time	Strategic	Quantitative
		Customer satisfaction	Strategic, tactical, and operational	Qualitative
		Effectiveness of distribution planning schedules	Tactical	Qualitative
		Level of customer perceived value of product	Strategic	
		Customer query time	Strategic	Quantitative
		Quality of delivery documentation	Operational	Qualitative
		Range of product and services	Strategic	Quantitative
3	Internal business operation perspective	Flexibility to meet particular customer needs	Strategic	Qualitative
		Total supply chain cycle time	Strategic	Quantitative
		Frequency of delivery	Operational	Qualitative
		Level of supplier's defect-free deliveries	Strategic	Qualitative
		Product development cycle time	Tactical	Quantitative
		Capacity utilization	Operational	Quantitative
		Planned process cycle time	Tactical	Quantitative

TABLE 4.2 *(Continued)*

Sr. No.	Balanced scorecard perspectives	Identified measures	Decision-making level	Nature of measures (quantitative/qualitative)
		Efficiency of purchase order cycle time	Operational	Quantitative
		Total cash flow time	Strategic	Quantitative
4	Innovation and learning perspective	Buyer–supplier partnership level	Strategic	Qualitative
		Accuracy of forecasting	Tactical	Qualitative
		Order entry method	Tactical	Qualitative
		Employee satisfaction and skill orientation	Strategic, tactical, and operational	Qualitative
		Supplier's booking procedures	Tactical	Qualitative
		Range of products/services	Strategic	Qualitative
		Supplier assistance in solving technical problems	Tactical	Qualitative
		Supplier ability to respond to quality problems	Tactical	Qualitative
		Delivery reliability	Operational	Qualitative

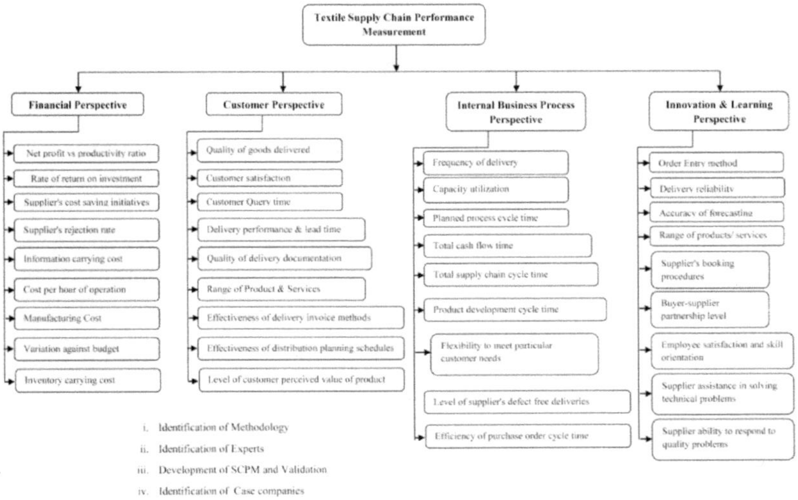

FIGURE 4.2 Supply chain performance measurement framework after literature review.

4.5.3 APPLICATION OF THE DELPHI TECHNIQUE FOR EXPLORING ISSUES FROM FIRST ROUND OF ONE-TO-ONE IN-DEPTH INTERVIEWS

As mentioned earlier in chapter 3, first step of survey questionnaire of the Delphi technique is modified by one-to-one in-depth interviews. The contents of these interviews are simple questions pertaining to general scene of the textile industry, supply chain management in the textile industry, supply chain performance measurement issues, and challenges and opportunities in Indian textile industry. Questions are highlighted as listed below and subsequent synthesized answer immediately following the question:

1. What is textile industry comprised of? Different processes of textile industry.
   Experts explained that textile is composed of various processes, such as ginning, spinning, weaving, processing/finishing, and further to different application-based manufacturing, that is, garment, home furnishing, and linings. But a major impact is created till fabric/cloth manufacturing. The industry is scattered with multiple units at different locations with high complicated chain.
2. What is industry's philosophy on SCM implementation?
   Interviewees replied that even though the concept of SCM is not new to manufacturing environment, the need is felt by every organization including manufacturing, service, and hospitality sector. As far, textile and clothing industries are concerned, SCM has not got much attention. The industries having integration with foreign counterparts for their business are much aware about the application of SCM. While, small and medium-sized enterprises are started very late to visualize the philosophy of SCM owing to intense global and local competitive environment. Genuine reasons for adopting SCM by the textile industry are as follows:
   i. Competitive pressure and insufficient profit gain for reducing prices.
   ii. Efficient utilization of financial assets to leverage market with cost-effective product.

iii. Establishing effective and friendly procurement strategies leading to supplier relationship management (SRM).
iv. Efficiently utilizing production facilities, thereby enhancing manufacturing efficiencies, resulting in lead time reduction, waste reduction, satisfying demand, etc.
v. Satisfying customer by distribution of manufactured product/ services in right quantity of right quality at right place and within right time (notion of SCM). Also providing better customer service.
vi. Enhanced integration of supply chain partners and effectively monitoring of supply chain cyclic processes such as procurement, production, and distribution.

3. What is state of evolution of performance evaluation activities among the firms to measure supply chain? Existing state, level of evolving the current state of art, and expectation in coming future. Please comment.

   Experts have agreed for the existence of measurement system among the firms, but not on comprehensive level. The different measurement approaches are such as conventional calculation of financial returns as compared to invested, activity-based costing (ABC), performance-based costing (PBC), etc. The approaches are not sufficient enough to measure the performance of supply chain management in textile/clothing industry. Evolving of supply chain performance measurement system is on progressive track as far as textile industries are concerned and drastic improvement is expected in future. Pinpointed remarks are as follows:

   i. The current state of art for supply chain performance measurement is evolving rapidly, but the issues in textile industry are how to measure the performance comprehensively by involving the nonfinancial aspects along with financial.
   ii. Other aspect is that many organizations feel how to involve the cross-functional approach. It means how to integrate supply chain partners/supply chain cyclic processes.
   iii. Identification and selection of measures should be based on integration among firms.
   iv. Concrete, integrated, proactive, and comprehensive measurement system will be evolved in near future.

v. Approaches will differ among the larger, small, and medium-sized enterprise.

4. Factors restricting the integration and orientation of the supply chain processes. Please elaborate.

Existing state of affairs, commitment from scattered unit structure, and communication capability were the prominent factors which obstruct the integration and orientation of supply chain performance measurement system. Few precise remarks are:

i. Enterprise structure and other associated issues such as commitment from top management, lack of sufficient infrastructure, unavailability of skilled manpower, and level of understanding between collaborating partners.
ii. Textile industry is full of technological obsolescence; hence, available IT infrastructure is obsolete and costly to upgrade. The larger players do not have this constraint; however, information sharing is still an issue in these enterprises.
iii. Ineffective selection of performance measures to enhancement in information shared used.
iv. Lack of functional/cyclic process-based performance measures for designing the strategy of an organization.
v. Envelop of performance measurement framework for textile supply chain.

5. What areas/aspects do you think that companies must look as important for effective and efficient measurement of supply chain performance?

Basically, it is difficult to design the most effective performance measurement approaches and thereby selection of such areas/aspects. Since a decade, it is in practice to use various industry-oriented aspects. But they are good for an enterprise, not for supply chain. Some precise comments are as follows:

i. The textile industry in India is facing global economic competitiveness, so in order to sustain in such globalization, economic stability is a very important aspect/perspective.
ii. Textile supply chain is very fragmented and complex in nature. It needs to serve the market by producing a variety of fabric/clothing patterns. For this, efficiency of internal business process is also a major area to focus.

   iii. Outdated technology leads the industry to sit back and face the loss of market share; hence, incorporating the innovative and learning environment would result in extracting reasonably good outputs.
   iv. Ultimate outcome of an industry is customer satisfaction. Fulfilling the needs of customer and solving their queries would result in achieving greater customer satisfaction.

6. What can be the key functions/activities being utilized to monitor the performance of supply chain in textile industries?
   With respect to textile industry, the operating units are so scattered, hence the supply chain becomes very complex. As a result of which performance measurement of textile supply chain is a cumbersome task. To facilitate this, precise comments are as follows:
   i. Consideration of activities in industries will not be sufficient.
   ii. In case of textile supply chain, cyclic processes can be considered to motivate the integration of supply chain partners such as ginning units, spinning units, etc.
   iii. Supply chain cyclic processes such as procurement, production, and distribution can be employed to monitor the supply chain performance.
   iv. Critical comment is made on the framework put forth in front of experts that merely identifying and selecting the performance measures under various BSC perspectives would not be sufficient; rather the important matter is to find the impact of these performance measures on various cyclic processes so as to realize the behavior of supply chain as a whole.
   v. Further modification is suggested in framework in terms of involving the supply chain cyclic processes, which is reflected in Figure 4.2.

7. Which performance attributes do you find most suitable to evaluate the performance of supply chain in textile and clothing industries?
   Experts assert that too many number of performance measures are not appropriate for effective evaluation of supply chain performance. Rather it creates ambiguity. Whereas, performance measures under each performance perspective relatively small in

number, yet clear and can be understandable by supervisor in the textile industry would be convenient.

Another important issue that should be kept in mind is selected performance measures must envelop the essence of whole supply chain. The suggested list of performance metrics/measures is as follows: (few of the performance measures were not remained the part of framework, whereas, performance measures with bold letter indicates that they are added by brainstorming exercise with expert and not emerged from literature review);

- **Financial perspective**
  - i. Rate of return on investment.
  - ii. Supplier cost saving initiative.
  - iii. Information carrying cost.
  - iv. Cost per hour of operation.
  - v. Inventory carrying cost: When this performance measure was decided to include, certain query raised in our mind that inventory carrying cost in process is worth? All experts explained the phenomena of inventory carrying cost as when raw material enters into spinning units, they are in the form of "BALES," block of raw material (cotton). Not all the raw material is made to process. Then the yarn is kept in the form of spindles and move on to weaving unit for further work to convert into fabric. In this way, inventory in the form of Bales, Spindles, Fabric blocks (may be limited, rest may be moved forward but at once) are kept in factory and carried as when required. This affects the coordination between cyclic processes of supply chain. Hence, "inventory carrying cost" is suggested.
  - vi. Manufacturing cost: The process of making textile fabric is very long and complex. The raw material has to pass through various numbers of operations right from cleaning in ginning, spinning, combing, recombing, weaving, processing, and finishing. At any moment, any small error can lead to rework/ even scrap, which ultimately contributes to cost of manufacturing. Because of this, fabric/cloth is made from cotton and mixing of other types of adhesive material. The quality of fabric depends on its strength and finish. Hence, "manufacturing cost" is suggested.

- **Customer perspective**
  i. Quality of goods delivered.
  ii. Customer satisfaction.
  iii. Customer query time.
  iv. Range of products and services.
  v. Delivery performance and lead time.
  vi. Effectiveness of distribution planning schedules.
- **Internal business process perspective**
  i. Capacity utilization.
  ii. Product development cycle time.
  iii. Planned process cycle time.
  iv. Flexibility of meeting particular customer needs.
  v. Total supply chain cycle time.
  vi. Extents of cooperation to improve quality: As a matter of fact, the textile supply chain is very complex and fragmented. Hence, effective communication and coordination among supply chain partners and also in the cyclic processes should be well maintained, so that the quality attainment is possible. Hence, "extent of cooperation to improve quality" is suggested.
- **Innovation and learning perspective**
  i. Buyer–supplier partnership level
  ii. Accuracy of forecasting
  iii. Employee satisfaction and skill orientation
  iv. Use of information technology (IT) and knowledge management (KM) concepts: The conventional way of maintaining data and sharing is obsolete and restricted to office files only. Estimating the quantity of raw material used and number of desired finished products manufactured can be achieved through this. It is very important in such fragmented supply chain to incorporate the application of sophisticated information sharing and enhanced knowledge management concepts. It has specific examples of online monitoring system across the supply chain partners, using high-end ERP packages; hence, it is suggested.
  v. Use of quality engineering (QE) and quality management techniques: Quality consciousness is very important in industry and across the supply chain. Quality management systems are

in implementation practice, but quality engineering concepts such as theory of constraints (TOC), Six Sigma, Taguchi, etc. would be helpful in evaluating the supply chain quality as a whole.

8. Please comment on the focus of this research study in the field of supply chain management and textile industry.
   Respondents with surplus knowledge of operation management in the interview were thinking about the requirement of supply chain performance measurement, whereas, experts from core textile field were more inclined toward the orientation of the textile industry. Still some clear responses are noted down;
   i. Performance measurement system for textile supply chain must be unambiguously defined and designed.
   ii. Probability of data collection from industry need to be very carefully monitored.
   iii. Merely identifying the performance measures and putting in a framework would not be sufficient, rather their relative impact on each other along with their impact on the supply chain cyclic processes must be evaluated using suitable method, in order to find the impact on overall supply chain performance.

4.5.4 OVERALL CONTRIBUTION OF FIRST ROUND OF INTERVIEWS

The contribution of first-level one-to-one in-depth interviews was remarkable. It provided direction for the research study. Subject matter experts in the field of textile industry with vast experience and adequate knowledge of supply chain management have consented that there is a need for such kind of exploratory studies. They highlighted the requirement of the philosophical shift of industry-oriented measurement system to measurement of supply chain cyclic processes connecting various enterprises of textile supply chain. Experts have agreed upon the concept and thinking of supply chain management and its performance measurement system in the Indian textile industry, resulting in the future success of textile businesses in India. They further added that the lack of implementation understanding hinders the concept of supply chain performance measurement in the textile industry in developing nation like India. Factors affecting implementation have been identified by experts. They also proposed supply chain performance perspective and associated criteria use to gauge the performance of textile supply chain. The important suggestions have been given by experts to modify the two-level hierarchical frameworks into three-level hierarchical structures. They suggested adding the supply chain cyclic processes at the third level in hierarchy, in order to evaluate the performance of whole supply chain. This first level of in-depth interview interaction showed that the supply chain performance measurement framework is critical for textile industry, as it requires more practical view in reality. Supply chain performance measurement system considering different levels consisting of overall goal, performance perspective as criteria, performance measures as subcriteria, and supply chain cyclic processes as alternative would be realistic and can be submitted for further investigation.

4.6 PURPOSE OF SECOND LEVEL OF DELPHI METHOD

The revised supply chain performance measurement framework, after incorporating the suggestions received from the experts, was presented to the experts again during the second round of the Delphi survey. The model is shown in Figure 4.2. The framework is identical to earlier framework as developed from the literature, whereas, depending on the suggestions received from experts during the first level of in-depth interview, revised framework is presented in Figure 4.3 after 1st and 2nd round of Delphi method. Further, experts were asked to consent to their agreement on the existence of identified parameters for the framework using the Likert scale from "strongly disagree" to "strongly agree"; for finalizing the element at various hierarchical levels as mentioned below.

i. Overall textile supply chain performance as goal/objective.
ii. BSC perspective as criteria to measure performance.
iii. Various performance measures as subcriteria under each BSC perspective.
iv. Supply chain cyclic processes as alternative to rate the performance of measures.

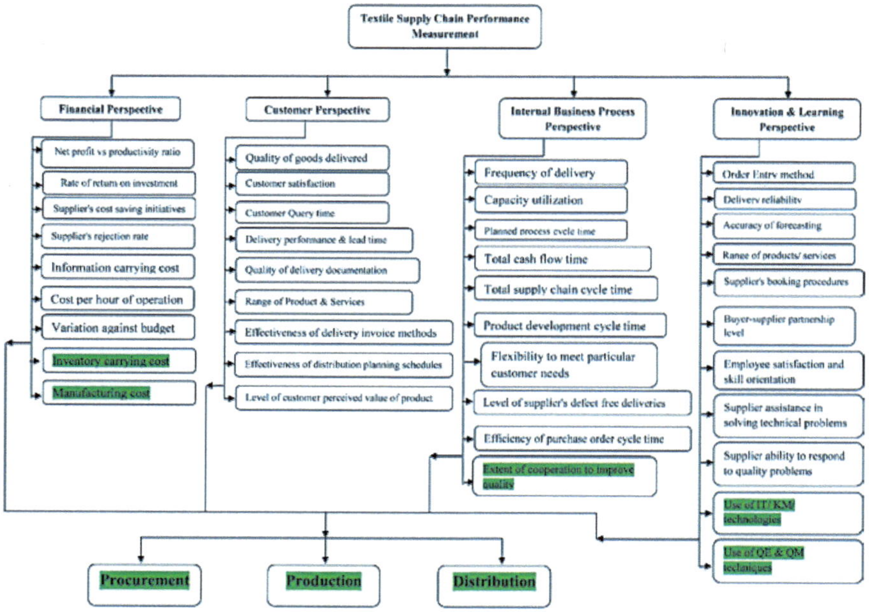

FIGURE 4.3 Revised SCPMS framework for textile supply chain (after 1st and 2nd stage).

4.6.1 RESULTS OF SECOND LEVEL OF DELPHI METHOD

Analyzing the Delphi method based on second level, results are shown in Tables 4.2 and 4.3. From Table 4.3, first-level criteria (BSC perspective) were analyzed by consent of experts. Coefficient of variation was employed to judge whether the experts have consensus with the available performance perspective as criteria. In practice, expert responses were supposed to be in consensus, if (COV<0.15) (Lin, 2003). It can be calculated as standard deviation/average value. Whereas in Table 4.3, the consensus on each and every performance measures as subcriteria and Cronbach's alpha was determined for each performance perspective/construct to examine the reliability. Reliability estimates for all the performance perspectives were found to be greater than 0.7, providing evidences for the internal consistency reached for the items (Hair et al., 1995).

TABLE 4.3 Second-level Delphi Results: First-level Performance Criteria Consensus.

Sr. No.	BSC perspective as performance criteria	Mean	Standard deviation	Coefficient of variation
1	Financial perspective	4.5	0.547	0.125
2	Customer perspective	4.3	0.523	0.118
3	Internal business process perspective	4.7	0.443	0.088
4	Innovation and learning perspective	4.2	0.410	0.079

As per the results obtained, coefficients of variation for all the BSC perspectives as first-level performance criteria are less than 0.15, which shows that experts have consensus for all the perspectives considered as first-level criteria for supply chain performance measurement of textile supply chain.

TABLE 4.4 Coefficient of Variation for Second-level Criteria and Cronbach's Alpha for Performance Constructs After Second Round of Delphi Method.

Sr. No.	Performance measurement criteria (BSC perspectives)	Performance subcriteria	Mean	COV	Cronbach's alpha
1	Financial perspective	Net profit versus productivity ratio	3.38	0.36	0.8253
		Rate of return on investment	4.82	0.093	
		Supplier's cost-saving initiatives	4.5	0.108	
		Supplier's rejection rate	3.15	0.408	
		Information carrying cost	4.24	0.106	
		Inventory carrying cost	4.6	0.092	
		Cost per hour of operation	4.5	0.109	
		Manufacturing cost	4.63	0.111	
		Variation against budget	3.08	0.333	

TABLE 4.4 *(Continued)*

Sr. No.	Performance measurement criteria (BSC perspectives)	Performance subcriteria	Mean	COV	Cronbach's alpha
2	Customer perspective	Quality of goods delivered	4.6	0.109	0.8497
		Customer satisfaction	4.59	0.111	
		Customer query time	4.22	0.106	
		Delivery performance and lead time	4.9	0.093	
		Quality of delivery documentation	2.7	0.298	
		Range of products and services	2.65	0.295	
		Effectiveness of delivery invoice methods	2.7	0.298	
		Effectiveness of distribution planning schedules	4.58	0.111	
		Level of customer perceived value of product	3.25	0.407	
3	Internal business process perspective	Frequency of delivery	3.4	0.335	0.8025
		Capacity utilization	4.8	0.093	
		Planned process cycle time	4.6	0.111	
		Total cash flow time	3.6	0.316	
		Total supply chain cycle time	4.6	0.111	
		Product development cycle time	4.4	0.120	
		Flexibility to meet particular customer needs	4.4	0.120	
		Level of supplier's defect-free deliveries	3	0.408	
		Extent of cooperation to improve quality	4.8	0.093	

TABLE 4.4 *(Continued)*

Sr. No.	Performance measurement criteria (BSC perspectives)	Performance subcriteria	Mean	COV	Cronbach's alpha
4	Innovation and learning perspective	Order entry method	2.2	0.280	0.7885
		Delivery reliability	2.6	0.319	
		Accuracy of forecasting	4.8	0.093	
		Range of products/services	2	0.5	
		Supplier's booking procedures	2.4	0.475	
		Buyer–supplier partnership level	4.8	0.093	
		Employee satisfaction and skill orientation	4.8	0.093	
		Supplier assistance in solving technical problems	2.8	0.296	
		Supplier ability to respond to quality problems	3	0.408	
		Use of IT/KM technologies	4.4	0.120	
		Use of QE and QM techniques	4.8	0.093	

The values of Cronbach's alpha for all the performance perspectives appeared to be greater than 0.7 and hence satisfactory. This results in verifying that performance measures not consented as not suitable for textile supply chain (coefficient of variation greater than 0.15) and those suitable to framework (COV less than 0.15) have consensus of expert and are reliable, as the reliability index for these performance perspectives has a value more than 0.7.

During interaction with the experts in the first round of interviews, precise comments have been received that the issues and practices pertaining to supply chain management and supply chain performance measurement differ with respect to size of the industry. Hence, while discussing about the focus of the proposed research study, a proposal has been kept by us to inspect the practices of SCM and SCPM in two different types of textile companies, which was encouraged by experts.

4.7 CONCLUSION OF THIRD LEVEL OF DELPHI METHOD

After executing the second level of Delphi method, experts again contacted telephonically and discussed for the consistency in the achieved result regarding deletion of some of the dissented performance measures, as observed by the coefficient of variation calculated for them individually and reliability of the associated higher level performance perspective. Experts have been in contact this time to verify the suitability level of these performance measures for textile supply chain.

At the end of the third level of Delphi method, the final supply chain performance measurement framework is developed as shown in Figure 4.4. As mentioned earlier, Figure 4.2 depicts SCPMS framework for the textile industry developed as per the observation from the literature review and with the understanding level of the author. Four performance attributes utilizing the benefits of BSC as strategic performance measurement system are identified and presented. At the end of first level of in-depth interviews of experts, earlier framework is revised as shown in Figure 4.3. Four new performance measures along with inclusion of supply chain cyclic processes to complete the hierarchical structure were suggested by experts. After second and third level of discussion, SCPMS for the textile and clothing industry is developed finally as shown in Figure 4.4. The original perspectives of BSC were retained as criteria with addition of four performance measures in existing ones as subcriteria and supply chain cyclic processes as alternative, thereby giving complete hierarchical structure to the framework.

4.8 IDENTIFICATION OF CASE COMPANIES FOR IMPLEMENTATION

For this purpose, we selected two large players and two small and medium enterprises of textile/clothing industry in our region. The identification of all four industries is from Maharashtra, as it provides justification with statics that Maharashtra is second in India with 1811 factories with cluster at multiple locations including Nagpur having annual turnover more than Rs. 1500 crores with variety of powerlooms and readymade garment manufacturers (CCI and CRISIL report). In all, four industries were selected (two large companies and two small and medium-sized enterprises). The names of the companies are kept anonymous as per the research ethics and also assured the industry personnel that the information sought (in form of ranking) would be used for academic purposes only.

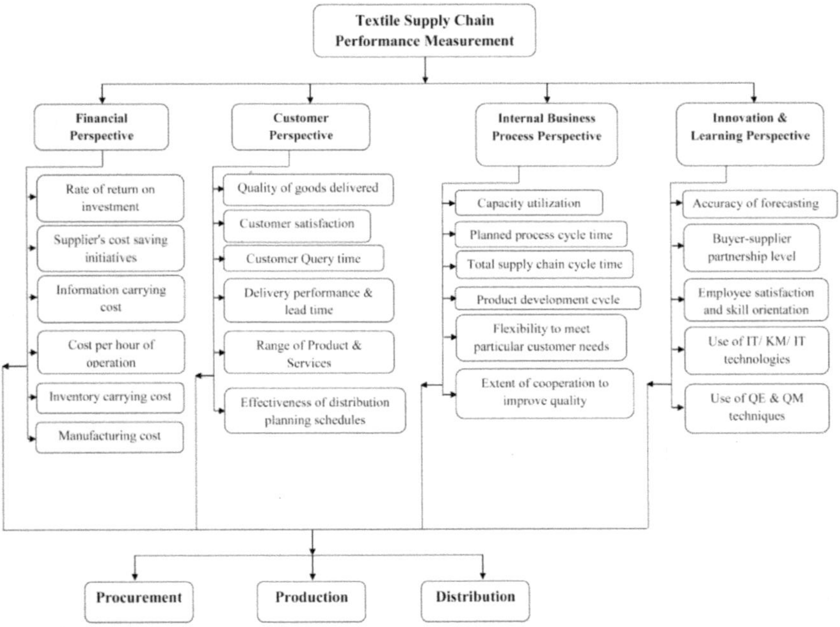

FIGURE 4.4 Final SCPMS framework for textile supply chain.

Along with the criteria mentioned earlier for selection of case companies, selection is influenced by practical accessibility of companies. Hence, we selected the companies located in our region.

4.8.1 PROFILES OF THE INDUSTRIES

- *Case Company 1*:
  The company is a part of larger group. The firm is entirely owned by Indians. The firm is producing almost everything,

especially suiting and shirting, but at different locations in the country. The company is almost more than 80 years old and having more than Rs. 1500 crores as annual turnover. The case company has more than 3500 employees working to produce high quality textured fabric for multiples usages. The company is branded and having more than 3000 retail outlets in the country and abroad also. The company used to export their finished fabric and yarn as well to countries like Italy, US, UK, Africa, etc.

- *Case Company 2*:
  The company belongs to a group which is almost 50 years old. The company was earlier a substituent of state textile corporation, but after collapse of the government corporation, it was owned by a group being one of the leaders in the field of textile/clothing industry. The company has almost 3000 employees and approximately more than Rs. 1000 crores annual turnover. The company produces high quality yarn and multiquality fabric. It exports products to European countries and Africa. The company's fabric is suitable for suiting–shirting, bed-lining, and home furnishing purposes.
- *Case Company 3*:
  The firm is 25 years old and having around 1200 employees. The company serves the Indian brand manufacturer with yarn and unprocessed fabric. Again, it had been a part of state's textile corporation but later owned by firm having various textile units in Gujarat with abundant experience relevant to the field. The company is having around Rs. 150–200 crores annual turnovers.
- *Case Company 4*:
  The firm is almost 20 years old and having around 800 employees. The company serves the Indian brand manufacturer with yarn and unprocessed fabric, which is used for manufacturing ethnic bed-lining, curtains, cushions, etc. The company is owned by a group having four spinning units (Soot-Girni). The company is having around Rs. 150–200 crore annual turnovers.

KEYWORDS

- **supply chain performance measurement**
- **textile industry**
- **framework development**
- **Delphi method**
- **key performance indicators**
- **performance measurement system**

CHAPTER 5

VALIDATION OF A DEVELOPED SCPM FRAMEWORK

ABSTRACT

This chapter provides an overview of the implementation of the supply chain performance measurement (SCPM) framework in case companies using analytical hierarchy process (AHP) methodology. It summarizes the implementation methodology, the analysis of data using AHP, and the application of AHP to the case companies. The implementation of the SCPM framework in case companies involves the adoption of the AHP methodology, a structured decision-making technique. AHP allows for a systematic and objective evaluation of different factors and criteria involved in SCPM. The implementation methodology consists of several steps, including the collection and analysis of relevant data, the identification of performance criteria, the establishment of hierarchy and weights for criteria, and the evaluation of alternatives. This methodology ensures a comprehensive and structured approach to implementing the SCPM framework. The analysis of data using AHP enables case companies to assess the relative importance of different criteria and factors in their SC performance. By assigning weights to criteria based on expert opinions and data analysis, companies can prioritize and focus on key areas that have the most significant impact on overall performance. The application of AHP to case companies involves applying the decision-making process to evaluate and compare various alternatives and strategies for improving SC performance. AHP provides a quantitative and transparent approach, enabling companies to make informed decisions based on rigorous analysis and criteria prioritization.

Supply Chain Performance Measurement in Textile Enterprises. Pranav C. Charkha, PhD, Santosh B. Jaju, PhD & Prasenjit Chatterjee, PhD (Authors.)

5.1 IMPLEMENTATION OF FRAMEWORK AND ANALYSIS FOR CASE COMPANIES USING ANALYTICAL HIERARCHY PROCESS

5.1.1 PURPOSE OF IMPLEMENTATION

Based on the suggestion of the experts during first and third rounds of interviews that supply chain performance measurement practices may vary with respect to type and size of industry. Hence, validation of framework and various performance measures in the different types of industry are needed and studied in this chapter.

- Instead of going for survey conduction, companies were identified and contacted.
- These companies are identified based on their involvement in manufacturing of various textile products.
- On request of industries, their names were kept anonymous.
- Four companies were consented and permitted to visit for interactions as listed below.
- Brief profile and features of case companies with its products mentioned in the bracket are as follows:

Case Company 1 (Fiber, fabric, apparel, and denim fabric)

- Belong to one of the largest brands of India.
- Expertise in producing, fabric, suiting-shirting, denim, etc.
- Group standing of almost more than 90 years.
- Annual turnover is about Rs. 1500 crores.
- Having around 3000 employees.
- Export the yarn and fabric to other countries like Italy, US, and UK, etc.

Case Company 2 (Fiber, multi-quality fabric for garment, and home furnishing)

- Earlier a constituent of state textile corporation.
- Now belongs to leading player of textile/clothing.
- Standing of almost more than 50 years.
- Annual turnover is about Rs. 1000 crores.
- Having almost 3000 employees.
- Produces high-quality yarn, multi-quality fabric useful for suiting as well for home furnishing.

- Export the products to European and African countries.

Case Company 3 (Fiber, unprocessed fabric for carpet production, and other utility products)

- Again were the part of MSTC.
- Later took over by Gujarat's leading textile and infrastructure development group.
- 25 years old company.
- Having around 900–1200 employees.
- Annual turnover approximately Rs. 100–150 crores.
- Company provides yarn and unprocessed fabric to manufacturer of carpet production and other utility textile products.

Case Company 4 (Fiber, synthetic fiber, fabric)

- Again were the part of MSTC.
- Now owned by KSL group.
- 20 years old company.
- Having around 1000 employees.
- Annual turnover approximately Rs. 100–150 crores.
- Company provides fiber, synthetic fiber, and rough fabric to manufacturer of sport utility product.

5.2 IMPLEMENTATION METHODOLOGY

As explained in Chapter 4, implementation of framework in case companies is suggested by expert and also it is required to validate the proposed framework with varied industrial contexts. This chapter identifies the applicability of developed framework and the importance of contained elements at different hierarchical levels.

The reason of this validation is

1. Suggestion given by experts in first and third rounds of the Delphi method.
2. To overcome the limitation and future scope suggested in studies of Allen (2008) and Pratabaraj (2012).

To study the practices of supply chain performance measurement and importance of performance measures of framework in identified textile case companies, respondents from different departments of an industry were identified to interact. The reasons for the identification of respondents are:

- ➢ To get a complete insight of industry on SCPM.
- ➢ To overcoming the limitation of the earlier study (Pratabaraj, 2012).

Respondents from 6 different departments along with representatives from top management are identified as shown in Table 4 in Chapter 4.

In total, 20 respondents identified from each company together make up 80 respondents.

All these respondents were supplied with the questionnaire prepared in the Delphi method and explained the importance of supply chain performance measurement in the textile industry. Also they were explained "how to fill the questionnaire."

They were asked to fill the questionnaire and responses were collected back in person. This task was carried out from June 2014 to December 2014.

Respondents were requested and reminded number of times by meeting in person. Out of 80, 75 questionnaires were returned.

A summary of completely filled questionnaire is mentioned in the figure as shown below:

Questionnaire submitted by respondents

('√' indicates questionnaire submitted by respondent, else not submitted)

Respondents/ Case Companies	Actors	Case Company 1	Case Company 2	Case Company 3	Case Company 4
Top Management	Managing Director	Not submitted	Not submitted	√	√
	General Manager/ Factory Manager	√	√	Not submitted	√
Commercial Department	Manager	√	√	√	√
	Asst. Manager	√	√	√	Not submitted
	Executive/ Supervisor	√	√	√	√
Procurement Department	Manager	√	√	√	√
	Asst. Manager	√	√	√	√
	Executive/ Supervisor	√	√	√	√
Production Department	Manager	√	√	√	√
	Asst. Manager	√	√	√	√
	Executive/ Supervisor	√	√	√	√
Distribution Department	Manager	√	√	√	√
	Asst. Manager	√	√	√	√
	Executive/ Supervisor	√	√	√	√
Quality Control Department	Manager	√	√	√	√
	Asst. Manager	√	√	√	√
	Executive/ Supervisor	√	√	√	√
Human Resource Department	Manager	√	√	Not submitted	√
	Asst. Manager	√	√	√	√
	Executive/ Supervisor	√	√	√	√

Data collected from Part A was organizational and personal information, from Part B understanding the concept of SCPM in the textile industry was revealed, from Part C verifying the development of SCPM framework was done, and from Part D pair-wise comparison of performance perspective as criteria and performance sub-criteria (factors) for estimating criticality of these measures toward supply chain performance and pair-wise comparison of supply chain cyclic operations with respect to individual performance measure toward overall supply chain performance of textile industry.

The data collected from Part D was the most useful data to establish the importance of framework and performance measures along with contribution of supply chain cyclic operation toward overall textile supply chain performance.

5.3 ANALYSIS OF DATA USING ANALYTICAL HIERARCHY PROCESS

The following chapter demonstrates the application of developed framework pertaining to performance measurement criteria and sub-criteria for measuring supply chain performance of the identified case companies to identify the level of criticality of performance measures to analyze their effect on selected textile industries supply chain performance by utilizing the algebra of analytic hierarchy process (AHP). The performance criteria, sub-criteria, alternative together in a framework were suggested and finalized for study and explained in earlier chapter (Chapter Four), whereas concept of analytical hierarchy process is explained in Chapter Three. The following section discusses details of performance measurement for supply chain management stranded on two major aspects: balanced scorecard performance perspective and supply chain cyclic processes. The results in this chapter are extracted by interviewing with six dissimilar types of departments (excluding representatives from top management) representing very macro activities performance under the three main supply chain cyclic processes in each identified case company. The respondents were categorized into 6 categories; respondents from procurement department representing relationship with suppliers, respondent from the production department giving the status of internal operation of an organization, respondent from distribution department representing the relationship with customers (wholesaler or retailer), respondents from the quality control department signifying the efforts done to impart various quality improvement initiatives/techniques, respondents from human resource department (HRD) giving an insight into what best and training/leaning adopted for employees to make them aware about evaluation of various types of methodologies in the context of supply chain management and respondents from commercial department handling issues like cost accounting, forecasting, etc.

For gaining more insight into the supply chain system at the strategic level, an additional session of interviews is conducted with top management (managing director, general manager). The structured questionnaire for taking the responses from the industry personnel is shown in Appendix A (Part A: General Information, Part B: SCM and SCPM in textile industry, Part C: Framework for SCPMS, and Part D: pair-wise comparison of criteria, sub-criteria, and alternatives). Hence, this chapter investigates crucial performance measurement criteria (four BSC perspective) and sub-criteria (associated indicator under each BSC perspective) and the level of impact/criticality of supply chain cyclic processes: procurement, production, and distribution. The output from the analysis of these two different types of case companies (Two larger- and two small- and medium-sized enterprises) facilitates the industries to formulate better strategies based on the evaluation supply chain performance under four BSC perspectives and using supply chain cyclic processes. The structured hierarchical representation is shown in Figure 5.1.

5.4 APPLICATION OF ANALYTICAL HIERARCHY PROCESS (AHP)

As stated earlier in Chapter Three, the analytical hierarchy process (AHP) is a structured process for representing complex problems hierarchically (Saaty, 1990). AHP facilitates modeling of unstructured problems (Jagdev et al., 2004). A structured hierarchy usually possesses three levels: top level in hierarchy represents objectives from a managerial standpoint, intermediate levels (second and third) explores the use of criteria and sub-criteria respectively which are used to gauge the intensity level of factors in application, and lowest level represents generally a list of alternatives (Bhagwat and Sharma, 2007b). AHP establishes the basic connectivity by segmenting a complex problem in smaller components and facilitates decision makers at managerial level using a sequence of pair-wise comparison judgments (can be examined and re-examined if needed) to explore the comparative strength or severity of impact of factors at different levels of hierarchy (Saaty, 1990). AHP exhibits the pair-wise comparison of factors/elements at the same hierarchical level (criteria, sub-criteria, and alternative). These subjective comparisons are transformed to numbers using the scale representing the importance of an element over another with respect to contributing to upper level in hierarchy (Saaty, 1980).

The proposed study elaborated the application of AHP to evaluate the performance of textile supply chain in developing countries like India, because of the following issues:

- The study in the field of textile supply chain management in India is at a very nascent stage.
- The earlier studies in India have concentrated on philosophy development and lack in providing the set of performance metrics and their intensity level of impact on supply chain performance.
- AHP provides an easily understood and compatible way of resolving complex problems.

It possesses the essence of multiple criteria decision-making technique which involves subjectivity and objectivity as well to facilitate decision-making approach (Dey, 2002). It involves active participation of various experts as decision makers in aggregating the consensus decision which provides decision makers a coherent basis making judgment for formulating a strategy. As stated by Saaty and Vargas (2013), AHP provides a theoretical ground for measurement to deal with quantifiable and intangible factors that can be considered for applications to various diversified areas, such as framing decision theory and conflict resolution.

Literature also suggests some critiques and debate in the use of AHP for decision-making. But, studies that exist justify the use of AHP in this framework, while it becomes mandatory to inspect the concerns raised by AHP challengers to satisfy their queries. Some studies disagreed that AHP has a concrete theoretical basis and a self-evident approach as compared to multi-attribute utility theory (MAUT) and is also queried for the legality of Saaty's perception; hence, it was a defective method in assessing decisions. It was then proved by Saaty (1990) using theorems that AHP relied on an axiomatic theory. In an attempt again Saaty (1990) fortified their point of view that the axioms of AHP are diverse as that of the traditional utility theory and are valid.

Later studies were conducted to provide the justification for selecting the AHP as a decision-making tool for supplier selection. The comparative methods used individually which include total cost approach, total cost of ownership (TCO), multiple attribute utility theory, data envelopment analysis, AHP, and relevant mathematical programming techniques. Studies suggested that TCO was a prominent approach, followed by AHP. Because of this finding, comparison held between TCO and AHP broadly. It suggested that AHP is useful and more comprehensive as well as a robust technique for decision makers to identify and gauge suppliers on various criteria, as compared to cost data only criteria in TCO (Ho et al., 2011).

Ho et al. (2010) reviewed articles on decision-making and evaluation. Various methods have been suggested for decision on supplier selection, such as using AHP, analytic network process (ANP), case-based reasoning (CBR), data envelopment analysis (DEA), fuzzy set theory, genetic algorithm (GA), mathematical programming, simple multi-attribute rating technique (SMART), and integrated techniques. According to Ho et al. (2010), DEA is most prevailing due to its robustness. It was also used to measure comparative efficiencies of uniform decision-making units (DMUs) based on quantitative data. Major limitations to DEA (Ho et al., 2011) are confusion pertaining to input and an output criterion, the second is ratings to qualitative criteria, and the third because of nature of DEA.

Hence, the second most popular stand-alone methodology is AHP, which is nowadays the most popular multi-criteria decision-making technique (Ho et al., 2010). It has wide application areas including education, engineering, finance, government, industry, management, manufacturing, personal, political, social, and sports since so many years (Ho, 2008). Its simplicity, ease of use, and great flexibility made it widely applicable. Moreover, its advantage is to deliver consistent decision with respect to multiple qualitative and quantitative factors. It is one of its inherent and special characteristics to extract consistency verification report index. It is used to avoid any inconsistency during personal or subjective decisions (Ho et al., 2011).

Hence, AHP is the most prominent tool for establishing priorities and selecting the best factor by prioritizing the performance measurement in routine operation. Furthermore, the AHP has a benefit that it provides a quantitative hierarchical framework in designing performance measurement system. Our study projected the measurement using four perspective of BSC: financial perspective, customer perspective, internal business process perspective and innovation, and learning perspective, considering aspects of all stakeholders.

Supply chain performance measurement is governed by three supply chain cyclic processes, which are evaluated for identifying the impact level of each BSC performance perspective. Under each performance measurement perspective, there exist a list of performance measurement sub-criteria and it is categorized as a three-decision level. The ultimate objective is to select the most critical supply chain cyclic processes among the three finalized earlier. These supply chain cyclic processes provide actual integration between the two supply chain partners and established the real meaning of supply chain management. The selection of performance criteria as BSC perspective is made in such a way that it should give us the consideration of all the stakeholders in one integrated model, which not only consider the stakeholder but also encompass the entire supply chain.

The echeloned representation of a problem is as follows: The first step is the breakdown of the complex problem into levels of hierarchy. At the top level is the goal as overall supply chain performance. At the second level are the factors or criteria of measurement as four BSC perspective, which considers all stakeholder and framing the strategic performance measurement system. These performance perspectives depict the ultimate objective of an organization leading to attainment of aim and goal of, that is, overall performance measurement system and at the third level are the sub-criteria of performance measurement factors categorized under each performance perspective then at the fourth/last level are the alternatives in the form of supply chain cyclic processes which are responsible for overall supply chain of textile enterprise.

Elements at the same level of hierarchy are pair wise compared and relative impact is estimated for the objective they would be attributing (Bhagwat and Sharma, 2007b). Comparative ratings are converted into a matrix form, which consist of elements, compared with each other resulting in a generation of square matrix. It has an identical number of rows and columns along with eigenvectors and eigen values (Bhagwat and Sharma, 2007b). The purpose of the algebraic calculation is to deliver the relative importance of factors quantitatively. The parameters/attributes/factors receiving highest values should be concentrated for developing and deploying any future plan of action.

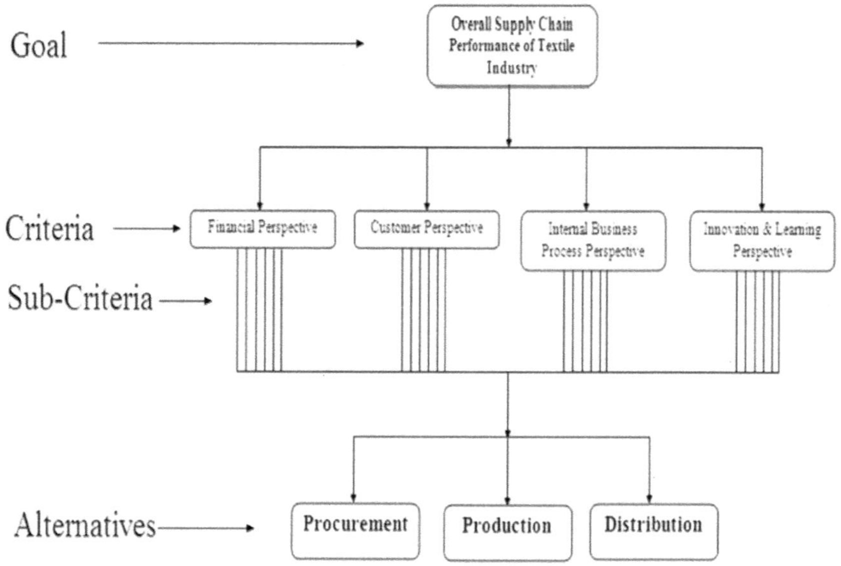

FIGURE 5.1 Hierarchical representation of framework for evaluating supply chain performance.

In this study for determining the supply chain performance measurement of textile industry, AHP decomposed problem into hierarchical levels, whereas a matrix is framed to assign the relative rating to criteria at the second level with respect to the overall goal and used to determine the relative importance. Further, the process is executed to construct matrices for executing pair-wise comparisons of each sub-criterion at the third level contributing to second-level elements. At last, similar matrices are constructed to exhibit pair-wise comparisons of each alternative in the fourth level with respect to the criteria, sub-criteria at second, third levels, respectively.

The calculation for priorities is done of second level by multiplying local priorities with priority of their respective criterion at the upper level and finally summing them for each element in a level as per criteria they affect. This is done to achieve local and global priorities of elements at that level and below evaluated by it as criterion and so on to lower level.

The tables of pair-wise comparison and weight calculation are shown for performance measurement factors for financial perspective, customer perspective, internal business process perspective, and innovation and learning perspective for each identified company, respectively. The results were discussed in detail in the next section/chapter.

5.5 APPLICATION OF AHP TO CASE COMPANIES

In executing the AHP method, we always execute the pair-wise comparison between elements at the same level with respect to element at higher level contributing to the goal of the study.

5.5.1 CASE COMPANY NO. 1

We first compare the performance criteria for estimating their relative weight.

TABLE 5.1 Tables of Pair-Wise Comparison and Weight Calculation for Performance Criteria (BSC Perspective).

Sr. No.	Performance criteria (BSC perspective)	1	2	3	4
1	Financial perspective	1	1/3	1/ 2.5	1/ 2
2	Customer perspective	3	1	1.5	2
3	Internal business perspective	2.5	1/1.5	1	1.5
4	Innovation and learning perspective	2	1/ 2	1/1.5	1

Sr. No.	Performance criteria (BSC perspective)	1	2	3	4	Weights
1	Financial perspective	0.117	0.133	0.112	0.1	0.116
2	Customer perspective	0.353	0.4	0.421	0.4	0.394
3	Internal business perspective	0.294	0.266	0.281	0.3	0.285
4	Innovation and learning perspective	0.235	0.2	0.187	0.2	0.205
						1.000

Max. Eigen Value (λ max) = 4.012

Consistency Index = 0.004 and Consistency Ratio = 0.0044.

TABLE 5.2 Tables of Pair-Wise Comparison and Weight Calculation for Performance Sub-Criteria under Financial Perspective.

Sr. No.	Sub-criteria under financial perspective	1	2	3	4	5	6
1	Rate of return on investment	1	1/3	1/ 2.5	1.5	1/ 2.5	2
2	Supplier's cost-saving initiatives	3	1	1.25	3	1.5	2.5
3	Information carrying cost	2.5	1/ 1.25	1	2.5	1.5	2
4	Cost per hour of operation	1/ 1.5	1/3	1/ 2.5	1	1/ 1.75	1/ 1.5
5	Inventory carrying cost	2.5	1/ 1.5	1/ 1.5	1.75	1	1.5
6	Manufacturing cost	1/2	1/ 2.5	1/2	1.5	1/1.5	1

Sr. No.	Performance sub-criteria under financial perspective	1	2	3	4	5	6	Weights
1	Rate of return on investment	0.098	0.094	0.095	0.151	0.071	0.207	0.116
2	Supplier's cost-saving initiatives	0.295	0.298	0.296	0.301	0.266	0.258	0.278
3	Information carrying cost	0.246	0.238	0.237	0.251	0.266	0.207	0.235
4	Cost per hour of operation	0.065	0.099	0.095	0.1	0.101	0.068	0.085
5	Inventory carrying cost	0.246	0.199	0.158	0.175	0.177	0.155	0.182
6	Manufacturing cost	0.049	0.119	0.118	0.022	0.118	0.103	0.104
								1.000

Max. Eigen Value (λ max) = 6.028
Consistency Index = 0.0056 and Consistency Ratio = 0.0045.

TABLE 5.3 Tables of Pair-Wise Comparison and Weight Calculation for Performance Sub-Criteria under Customer Perspective.

Sr. No.	Sub-criteria under customer perspective	1	2	3	4	5	6
1	Quality of goods manufactured	1	1.5	1/3	2	1/ 1.5	1.75
2	Delivery performance and lead time	1/ 1.5	1	1/3	1.75	½	1.5
3	Customer satisfaction	3	3	1	3.25	1.5	2.5
4	Customer query time	1/2	1/1.75	1/ 3.25	1	1/ 2.5	1/ 1.5
5	Range of product and services	1.5	2	1/ 1.5	2.5	1	2
6	Effectiveness of distribution planning schedules	1/1.75	1/ 1.5	1/ 2.5	1.5	1/2	1

Sr. No.	Performance sub-criteria under customer perspective	1	2	3	4	5	6	Weights
1	Quality of goods manufactured	0.138	0.172	0.109	0.166	0.145	0.185	0.154
2	Delivery performance and lead time	0.092	0.114	0.109	0.146	0.109	0.159	0.125
3	Customer satisfaction	0.414	0.343	0.329	0.271	0.328	0.265	0.320
4	Customer query time	0.069	0.065	0.101	0.083	0.087	0.07	0.080
5	Range of product and services	0.207	0.228	0.219	0.208	0.219	0.212	0.217
6	Effectiveness of distribution planning schedules	0.078	0.076	0.132	0.125	0.109	0.106	0.104
								1.000

Max. Eigen Value (λ max) = 6.033

Consistency Index = 0.006 and Consistency Ratio = 0.0048.

TABLE 5.4 Tables of Pair-Wise Comparison and Weight Calculation for Performance Sub-Criteria under Internal Business Process Perspective.

Sr. No.	Sub-criteria under internal business process perspective	1	2	3	4	5	6
1	Capacity utilization	1	1/1.75	1/3	1/1.5	1/2	1/2.5
2	Planned process cycle time	1.75	1	1/ 2.5	1.5	1/ 1.5	1/2
3	Total supply chain cycle time	3	2.5	1	2.5	1.75	1.5
4	Product development cycle time	1.5	1/ 1.5	1/ 1.75	1	1/ 1.75	1/ 2.5
5	Flexibility to meet particular customer needs	2	1.5	1/ 1.75	1.75	1	1/ 1.5
6	Extent of cooperation to improve quality	2.5	2	1/ 1. 5	2.50	1.5	1

Sr. No.	Performance sub-criteria under internal business process perspective	1	2	3	4	5	6	Weights
1	Capacity utilization	0.085106	0.069	0.0940	0.067	0.083	0.089	0.082
2	Planned process cycle time	0.148936	0.124	0.1130	0.151	0.111	0.112	0.127
3	Total supply chain cycle time	0.255319	0.303	0.2820	0.252	0.292	0.336	0.290
4	Product development cycle time	0.12766	0.081	0.1610	0.1	0.095	0.089	0.101
5	Flexibility to meet particular customer needs	0.170213	0.181	0.1610	0.176	0.167	0.149	0.169
6	Extent of cooperation to improve quality	0.212766	0.243	0.1880	0.252	0.25	0.224	0.230
								1.000

Max. Eigen Value (λ max) = 6.0113

Consistency Index =0.0022 and Consistency Ratio= 0.0018.

TABLE 5.5 Tables of Pair-Wise Comparison and Weight Calculation for Performance Sub-Criteria under Innovation and Learning Perspective.

Sr. No.	Sub-criteria under innovation and learning perspective	1	2	3	4	5
1	Accuracy of forecasting	1	1/2.5	1.25	1/1.25	1/2
2	Buyer supplier partnership level	2.5	1	3	1.5	1.5
3	Employee satisfaction and skill orientation	1/1.25	1/3	1	1/1.75	1/2
4	Use of IT and KM technologies	1.25	1/1.5	1.75	1	1/1.5
5	Use of QE and QM techniques	2	1/1.5	2	1.5	1

Sr. No.	Performance sub-criteria innovation and learning perspective	1	2	3	4	5	Weights
1	Accuracy of forecasting	0.132	0.13	0.138	0.148	0.119	0.134
2	Buyer supplier partnership level	0.331	0.326	0.333	0.282	0.359	0.326
3	Employee satisfaction and skill orientation	0.106	0.108	0.111	0.106	0.119	0.110
4	Use of IT and KM technologies	0.165	0.217	0.194	0.186	0.159	0.185
5	Use of QE and QM techniques	0.265	0.217	0.222	0.279	0.239	0.245
							1.000

Max. Eigen Value (λ max) =5.033

Consistency Index = 0.008 and Consistency Ratio = 0.009.

Now, to find the impact of individual performance sub-criteria on the supply chain cyclic processes (alternatives in supply chain performance measurement framework).

Pair-wise comparison of alternatives with respect to performance criteria under financial perspective.

TABLE 5.6 Pair-Wise Comparison of Alternatives with Respect to Sub-Criteria (Rate of Return on Investment) for Analysis the Impact on Overall Supply Chain Performance.

Sr. No.	Supply chain cyclic processes (alternatives)	1	2	3
1	Procurement	1	1/2	1/ 1.5
2	Production	2	1	2
3	Distribution	1.5	1/2	1

Sr. No.	Supply chain cyclic processes (alternatives)	1	2	3	Weights
1	Procurement	0.226	0.25	0.182	0.217
2	Production	0.444	0.5	0.545	0.498
3	Distribution	0.333	0.25	0.273	0.285
					1.000

A: Sub-criteria: Rate of return on investment.
Max. Eigen Value (λ max) = 3.016
Consistency Index = 0.002 and Consistency Ratio = 0.0034.

TABLE 5.7 Pair-Wise Comparison of Alternatives With Respect to Sub-Criteria (Supplier's Cost-Saving Initiative) for Analysis the Impact on Overall Supply Chain Performance.

Sr. No.	Supply chain cyclic processes (alternatives)	1	2	3
1	Procurement	1	1.5	2
2	Production	1/1.5	1	1.5
3	Distribution	1/2	1/1.5	1

Sr. No.	Supply chain cyclic processes (alternatives)	1	2	3	Weights
1	Procurement	0.461	0.473	0.444	0.461
2	Production	0.307	0.315	0.333	0.319
3	Distribution	0.231	0.210	0.222	0.221
					1.000

B: Sub-criteria: Supplier's cost-saving initiatives.
Max. Eigen Value (λ max) = 3.006
Consistency Index = 0.003 and Consistency Ratio = 0.0051.

TABLE 5.8 Pair-Wise Comparison of Alternatives with Respect to Sub-Criteria (Information Carrying Cost) for Analysis of the Impact on Overall Supply Chain Performance.

Sr. No.	Supply chain cyclic processes (alternatives)	1	2	3
1	Procurement	1	1.5	1.25
2	Production	1/1.5	1	1.5
3	Distribution	1/1.25	1/1.5	1

Sr. No.	Supply chain cyclic processes (alternatives)	1	2	3	Weights
1	Procurement	0.405	0.474	0.333	0.406
2	Production	0.270	0.316	0.400	0.328
3	Distribution	0.324	0.210	0.267	0.266
					1.000

C: - Sub-criteria: Information carrying cost.
Max. Eigen Value (λ max) = 3.099
Consistency Index = 0.048 and Consistency Ratio = 0.08.

TABLE 5.9 Pair-Wise Comparison of Alternatives With Respect to Sub-Criteria (Cost Per Hour of Operation) for Analysis of the Impact on Overall Supply Chain Performance.

Sr. No.	Supply chain cyclic processes (alternatives)	1	2	3
1	Procurement	1	1/2	1/1.25
2	Production	2	1	2
3	Distribution	1.25	1/2	1

Sr. No.	Supply chain cyclic processes (alternatives)	1	2	3	Weights
1	Procurement	0.235	0.25	0.117	0.201
2	Production	0.470	0.5	0.588	0.519
3	Distribution	0.294	0.25	0.294	0.279
					1.000

D: - Sub-criteria: Cost per hour of operation.
Max. Eigen Value (λ max) = 3.035
Consistency Index = 0.017 and Consistency Ratio = 0.03.

TABLE 5.10 Pair-Wise Comparison of Alternatives With Respect to Sub-Criteria (Inventory Cost) for Analysis of the Impact on Overall Supply Chain Performance.

Sr. No.	Supply chain cyclic processes (alternatives)	1	2	3
1	Procurement	1	1/2	1.25
2	Production	2	1	1.75
3	Distribution	1/1.25	1/1.75	1

Sr. No.	Supply chain cyclic processes (alternatives)	1	2	3	Weights
1	Procurement	0.263	0.241	0.312	0.272
2	Production	0.526	0.483	0.437	0.483
3	Distribution	0.210	0.276	0.250	0.245
					1.000

E: Sub-criteria: Inventory carrying cost.
Max. Eigen Value (λ max) = 3.005
Consistency Index =0.0025 and Consistency Ratio= 0.0043.

TABLE 5.11 Pair-Wise Comparison of Alternatives With Respect to Sub-Criteria (Manufacturing Cost) for Analysis of the Impact on Overall Supply Chain Performance.

Sr. No.	Supply chain cyclic processes (alternatives)	1	2	3
1	Procurement	1	1/ 2.5	1
2	Production	2.5	1	2
3	Distribution	1	1/2	1

Sr. No.	Supply chain cyclic processes (alternatives)	1	2	3	Weights
1	Procurement	0.222	0.210	0.25	0.227
2	Production	0.555	0.526	0.5	0.528
3	Distribution	0.222	0.263	0.25	0.245
					1.000

F: Sub-criteria: Manufacturing cost.
Max. Eigen Value (λ max) = 3.002
Consistency Index =0.001 and Consistency Ratio= 0.0017.
Pair-wise comparison of alternatives with respect to performance criteria under customer perspective.

TABLE 5.12 Pair-Wise Comparison ofAlternatives with Respect to Sub-Criteria (Quality of Goods Manufactured) for Analysis of the Impact on Overall Supply Chain Performance.

Sr. No.	Supply chain cyclic processes (alternatives)	1	2	3
1	Procurement	1	1/2.5	1/1.5
2	Production	2.5	1	2.25
3	Distribution	1.5	1/2.25	1

Sr. No.	Supply chain cyclic processes (alternatives)	1	2	3	Weights
1	Procurement	0.2	0.217	0.170	0.196
2	Production	05	0.543	0.575	0.539
3	Distribution	0.3	0.240	0.255	0.265
					1.000

G: Sub-criteria: Quality of goods manufactured.
Max. Eigen Value (λ max) =3.098
Consistency Index = 0.049 and Consistency Ratio= 0.08.

TABLE 5.13 Pair-Wise Comparison of Alternatives With Respect to Sub-Criteria (Delivery Performance and Lead Time) for Analysis of the Impact on Overall Supply Chain Performance.

Sr. No.	Supply chain cyclic processes (alternatives)	1	2	3
1	Procurement	1	1/1.5	1/1.5
2	Production	1.5	1	2
3	Distribution	1.5	1/2	1

Sr. No.	Supply chain cyclic processes (alternatives)	1	2	3	Weights
1	Procurement	0.25	0.307	0.182	0.246
2	Production	0.375	0.462	0.545	0.461
3	Distribution	0.375	0.231	0.273	0.293
					1.000

H: Sub-criteria: Delivery performance and lead time.
Max. Eigen Value (λ max) = 3.001
Consistency Index = 0.0005 and Consistency Ratio=0.0008.

TABLE 5.14 Pair-Wise Comparison of Alternatives With Respect to Sub-Criteria (Customer Satisfaction) for Analysis of the Impact on Overall Supply Chain Performance.

Sr. No.	Supply chain cyclic processes (alternatives)	1	2	3
1	Procurement	1	1/1.5	1/1.25
2	Production	1.5	1	1.25
3	Distribution	1.25	1/1.25	1

Sr. No.	Supply chain cyclic processes (alternatives)	1	2	3	Weights
1	Procurement	0.267	0.270	0.262	0.266
2	Production	0.4	0.405	0.409	0.405
3	Distribution	0.333	0.324	0.327	0.329
					1.000

I: Sub-criteria: Customer satisfaction.
Max. Eigen Value (λ max) = 3.012
Consistency Index =0.006 and Consistency Ratio= 0.01.

TABLE 5.15 Pair-Wise Comparison of Alternatives With Respect to Sub-Criteria (Effectiveness of Distribution Planning Schedules) for Analysis of the Impact on Overall Supply Chain Performance.

Sr. No.	Supply chain cyclic processes (alternatives)	1	2	3
1	Procurement	1	1.25	1/2
2	Production	1/1.25	1	1/1.5
3	Distribution	2	1.5	1

Sr. No.	Supply chain cyclic processes (alternatives)	1	2	3	Weights
1	Procurement	0.263	0.333	0.231	0.275
2	Production	0.21	0.267	0.307	0.261
3	Distribution	0.526	0.400	0.461	0.464
					1.000

J: Sub-criteria: Effectiveness of distribution planning schedules.
Max. Eigen Value (λ max) = 3.002
Consistency Index = 0.001 and Consistency Ratio = 0.0017.

TABLE 5.16 Pair-Wise Comparison of Alternatives With Respect to Sub-Criteria (Customer Query Time) for Analysis of the Impact on Overall Supply Chain Performance.

Sr. No.	Supply chain cyclic processes (alternatives)	1	2	3
1	Procurement	1	1.25	1/1.25
2	Production	1/1.25	1	1/1.25
3	Distribution	1.25	1.25	1

Sr. No.	Supply chain cyclic processes (alternatives)	1	2	3	Weights
1	Procurement	0.327	0.357	0.307	0.331
2	Production	0.262	0.285	0.307	0.285
3	Distribution	0.409	0.357	0.385	0.384
					1.000

K: Sub-criteria: Customer query time.
Max. Eigen Value (λ max) = 3.018
Consistency Index = 0.009 and Consistency Ratio = 0.015.

TABLE 5.17 Pair-Wise Comparison of Alternatives With Respect to Sub-Criteria (Range of Product and Services) for Analysis of the Impact on Overall Supply Chain Performance.

Sr. No.	Supply chain cyclic processes (alternatives)	1	2	3
1	Procurement	1	1/2.5	2
2	Production	2.5	1	3
3	Distribution	½	1/3	1

Sr. No.	Supply chain cyclic processes (alternatives)	1	2	3	Weights
1	Procurement	0.25	0.231	0.333	0.270
2	Production	0.625	0.578	0.5	0.570
3	Distribution	0.125	0.193	0.167	0.160
					1.000

L: Sub-criteria: Range of product and services.
Max. Eigen Value (λ max) = 3.010
Consistency Index = 0.005 and Consistency Ratio= 0.008.
Pair-wise comparison of alternatives with respect to performance criteria under internal business process perspective.

TABLE 5.18 Pair-Wise Comparison of Alternatives With Respect to Sub-Criteria (Flexibility to Meet Particular Customer Needs) for Analysis of the Impact on Overall Supply Chain Performance.

Sr. No.	Supply chain cyclic processes (alternatives)	1	2	3
1	Procurement	1	1/2.7	1/1.8
2	Production	2.7	1	2.4
3	Distribution	1.8	1/2.4	1

Sr. No.	Supply chain cyclic processes (alternatives)	1	2	3	Weights
1	Procurement	0.327	0.234	0.253	0.263
2	Production	0.491	0.562	0.607	0.554
3	Distribution	0.182	0.208	0.141	0.184
					1.000

M: Sub-criteria: Flexibility to meet particular customer needs.
Max. Eigen Value (λ max) = 3.0261
Consistency Index = 0.013 and Consistency Ratio=0.022.

TABLE 5.19 Pair-Wise Comparison of Alternatives With Respect to Sub-Criteria (Total Supply Chain Cycle Time) for Analysis of the Impact on Overall Supply Chain Performance.

Sr. No.	Supply chain cyclic processes (alternatives)	1	2	3
1	Procurement	1	1/1.25	1.8
2	Production	1.25	1	2.2
3	Distribution	1/1.8	1/2.2	1

Sr. No.	Supply chain cyclic processes (alternatives)	1	2	3	Weights
1	Procurement	0.357	0.355	0.36	0.356
2	Production	0.446	0.444	0.44	0.444
3	Distribution	0.198	0.202	0.20	0.200
					1.000

N: Sub-criteria: Total supply chain cycle time.
Max. Eigen Value (λ max) = 3.011
Consistency Index = 0.005 and Consistency Ratio = 0.0094.

TABLE 5.20 Pair-Wise Comparison of Alternatives With Respect to Sub-Criteria (Product Development Cycle Time) for Analysis of the Impact on Overall Supply Chain Performance.

Sr. No.	Supply chain cyclic processes (alternatives)	1	2	3
1	Procurement	1	1/2.3	1/1.5
2	Production	2.3	1	2
3	Distribution	1.5	1/2	1

Sr. No.	Supply chain cyclic processes (alternatives)	1	2	3	Weights
1	Procurement	0.312	0.257	0.272	0.265
2	Production	0.479	0.515	0.545	0.526
3	Distribution	0.208	0.224	0.186	0.210
					1.000

O: Sub-criteria: Product development cycle time.
Max. Eigen Value (λ max) = 3.0010
Consistency Index = 0.0005 and Consistency Ratio = 0.00086.

TABLE 5.21 Pair-Wise Comparison of Alternatives With Respect to Sub-Criteria (Capacity Utilization) for Analysis of the Impact on Overall Supply Chain Performance.

Sr. No.	Supply chain cyclic processes (alternatives)	1	2	3
1	Procurement	1	1/3	1/1.5
2	Production	3	1	1.5
3	Distribution	1.5	1/1.5	1

Sr. No.	Supply chain cyclic processes (alternatives)	1	2	3	Weights
1	Procurement	0.182	0.167	0.210	0.186
2	Production	0.545	0.500	0.474	0.507
3	Distribution	0.273	0.333	0.316	0.307
					1.000

P: Sub-criteria: Capacity utilization.
Max. Eigen Value (λ max) = 3.0025
Consistency Index = 0.00125 and Consistency Ratio = 0.0021.

TABLE 5.22 Pair-Wise Comparison of Alternatives With Respect to Sub-Criteria (Planned Process Cycle Time) for Analysis of the Impact on Overall Supply Chain Performance.

Sr. No.	Supply chain cyclic processes (alternatives)	1	2	3
1	Procurement	1	1/1.25	1.25
2	Production	1.25	1	2
3	Distribution	1/1.25	½	1

Sr. No.	Supply chain cyclic processes (alternatives)	1	2	3	Weights
1	Procurement	0.327	0.347	0.294	0.322
2	Production	0.409	0.435	0.470	0.438
3	Distribution	0.262	0.217	0.235	0.238
					1.000

Q: - Sub-criteria: Planned process cycle time.
Max. Eigen Value (λ max) = 3.0136
Consistency Index = 0.0068 and Consistency Ratio = 0.0117.

TABLE 5.23 Pair-Wise Comparison of Alternatives With Respect to Sub-Criteria (Extent of Cooperation to Improve Quality) for Analysis of the Impact on Overall Supply Chain Performance.

Sr. No.	Supply chain cyclic processes (alternatives)	1	2	3
1	Procurement	1	1/2.75	1.7
2	Production	2.75	1	2.8
3	Distribution	1/1.7	1/ 2.8	1

Sr. No.	Supply chain cyclic processes (alternatives)	1	2	3	Weights
1	Procurement	0.230	0.211	0.309	0.250
2	Production	0.634	0.581	0.509	0.575
3	Distribution	0.135	0.207	0.182	0.175
					1.000

R: Sub-criteria: Extent of cooperation to improve quality.
Max. Eigen Value (λ max) = 3.0082
Consistency Index = 0.0041 and Consistency Ratio = 0.0070.
Pair-wise comparison of alternatives with respect to performance criteria under innovation and learning perspective.

TABLE 5.24 Pair-Wise Comparison of Alternatives With Respect to Sub-Criteria (Accuracy of Forecasting) for Analysis of the Impact on Overall Supply Chain Performance.

Sr. No.	Supply chain cyclic processes (alternatives)	1	2	3
1	Procurement	1	2.5	1.5
2	Production	1/2.5	1	1.25
3	Distribution	1/1.5	1/1.25	1

Sr. No.	Supply chain cyclic processes (alternatives)	1	2	3	Weights
1	Procurement	0.484	0.581	0.4	0.488
2	Production	0.193	0.232	0.333	0.253
3	Distribution	0.323	0.186	0.266	0.258
					1.000

S: Sub-criteria: Accuracy of forecasting.
Max. Eigen Value (λ max) = 3.0010
Consistency Index = 0.005 and Consistency Ratio = 0.0008.

TABLE 5.25 Pair-Wise Comparison of Alternatives With Respect to Sub-Criteria (Buyer Supplier Partnership Level) for Analysis of the Impact on Overall Supply Chain Performance.

Sr. No.	Supply chain cyclic processes (alternatives)	1	2	3
1	Procurement	1	3	1
2	Production	1/3	1	1.6
3	Distribution	1	1/1.6	1

Sr. No.	Supply chain cyclic processes (alternatives)	1	2	3	Weights
1	Procurement	0.429	0.649	0.278	0.456
2	Production	0.143	0.216	0.444	0.182
3	Distribution	0.429	0.135	0.278	0.362
					1.000

T: Sub-criteria: Buyer supplier partnership level.
Max. Eigen Value (λ max) = 3.0189
Consistency Index = 0.0094 and Consistency Ratio = 0.016.

TABLE 5.26 Pair-Wise Comparison of Alternatives With Respect to Sub-Criteria (Use of IT and KM Technologies) for Analysis of the Impact on Overall Supply Chain Performance.

Sr. No.	Supply chain cyclic processes (alternatives)	1	2	3
1	Procurement	1	1/1.5	1.25
2	Production	1.5	1	2
3	Distribution	1/1.25	1/2	1

Sr. No.	Supply chain cyclic processes (alternatives)	1	2	3	Weights
1	Procurement	0.303	0.307	0.294	0.301
2	Production	0.454	0.461	0.471	0.462
3	Distribution	0.242	0.231	0.235	0.253
					1.000

U: Sub-criteria: Use of IT and KM concepts.
Max. Eigen Value (λ max) = 3.0010
Consistency Index = 0.0005 and Consistency Ratio = 0.00086.

TABLE 5.27 Pair-Wise Comparison of Alternatives With Respect to Sub-Criteria (Use of QE and QM Techniques) for Analysis of the Impact on Overall Supply Chain Performance.

Sr. No.	Supply chain cyclic processes (alternatives)	1	2	3
1	Procurement	1	1/3	1.5
2	Production	3	1	3.2
3	Distribution	1/1.5	1/3.2	1

Sr. No.	Supply chain cyclic processes (alternatives)	1	2	3	Weights
1	Procurement	0.214	0.203	0.263	0.226
2	Production	0.643	0.608	0.561	0.603
3	Distribution	0.143	0.189	0.175	0.169
					1.000

V: Sub-criteria: Use of QE and QM techniques.
Max. Eigen Value (λ max) = 3.0024
Consistency Index = 0.0012 and Consistency Ratio = 0.00206.

TABLE 5.28 Pair-Wise Comparison of Alternatives With Respect to Sub-Criteria (Employee Satisfaction and Skill Orientation) for Analysis of the Impact on Overall Supply Chain Performance.

Sr. No.	Supply chain cyclic processes (alternatives)	1	2	3
1	Procurement	1	1/2	1.5
2	Production	2	1	2.2
3	Distribution	1/1.5	1/2.2	1

Sr. No.	Supply chain cyclic processes (alternatives)	1	2	3	Weights
1	Procurement	0.273	0.256	0.319	0.283
2	Production	0.545	0.513	0.468	0.508
3	Distribution	0.182	0.233	0.213	0.209
					1.000

W: Sub-criteria: Employee satisfaction and skill orientation.
Max. Eigen Value (λ max) = 3.0030
Consistency Index = 0.0015 and Consistency Ratio = 0.0025.

TABLE 5.29 Importance of Supply Chain Cyclic Processes With Respect to Overall Supply Chain Performance Measurement.

Sr. No.	Supply chain cyclic processes (alternatives)	Final weight
1	Procurement	0.302
2	Production	0.435
3	Distribution	0.263

TABLE 5.30 Appendix B: Comparative Priorities of Supply Chain Cyclic Processes With Respect to Various Performance Measures (Sub-Criteria).

	1	2	3	4	5	6
	Financial Perspective					
Level of Sub-criteria	St	Tt	Op	Op	Op	Op
Supply Chain Cyclic Processes	A	B	C	D	E	F
Procurement	0.317	**0.461**	**0.406**	0.301	0.272	0.227
Production	**0.498**	0.319	0.328	**0.519**	**0.483**	**0.528**
Distribution	0.215	0.221	0.266	0.279	0.245	0.245
Max. Eigen Value (λ max)	3.016	3.006	3.099	3.035	3.005	3.002
Consistency Index	0.002	0.003	0.048	0.017	0.0025	0.001
Consistency Ratio	0.0034	0.0051	0.08	0.03	0.0043	0.0017

	7	8	9	10	11	12
	Customer Perspective					
Level of Sub-criteria	Op	St	S, T, O	Tt	St	St
Supply Chain Cyclic Processes	G	H	I	J	K	L
Procurement	0.196	0.246	0.266	0.275	0.331	0.27
Production	**0.539**	**0.461**	**0.405**	0.261	0.285	**0.57**
Distribution	0.265	0.293	0.329	**0.464**	**0.384**	0.16
Max. Eigen Value (λ max)	3.098	3.001	3.012	3.002	3.018	3.01
Consistency Index	0.049	0.0005	0.006	0.001	0.009	0.005
Consistency Ratio	0.08	0.0008	0.01	0.0017	0.015	0.008

	13	14	15	16	17	18
	Internal Business Process Perspective					
Level of Sub-criteria	St	St	Tt	Op	Tt	S, T, O
Supply Chain Cyclic Processes	M	N	O	P	Q	R
Procurement	0.263	0.356	0.265	0.185	0.322	0.25
Production	**0.554**	**0.444**	**0.526**	**0.507**	**0.438**	**0.575**
Distribution	0.184	0.2	0.21	0.307	0.238	0.175
Max. Eigen Value (λ max)	3.0261	3.011	3.001	3.0025	3.0136	3.0082
Consistency Index	0.013	0.005	0.0005	0.00125	0.0068	0.0041
Consistency Ratio	0.022	0.0094	0.00086	0.0021	0.0117	0.007

	19	20	21	22	23
	Innovation & Learning Perspective				
Level of Sub-criteria	St, Tt	St	S, T, O	S, T, O	S, T, O
Supply Chain Cyclic Processes	S	T	U	V	W
Procurement	**0.488**	**0.456**	0.301	0.226	0.283
Production	0.253	0.182	**0.462**	**0.603**	**0.508**
Distribution	0.258	0.362	0.253	0.169	0.209
Max. Eigen Value (λ max)	3.001	3.0189	3.001	3.0024	3.003
Consistency Index	0.005	0.0094	0.0005	0.0012	0.0015
Consistency Ratio	0.0008	0.016	0.00086	0.00206	0.0025

5.5.2 *CASE COMPANY NO. 2*

We first compare the performance criteria for estimating their relative weight.

TABLE 5.31 Tables of Pair-Wise Comparison and Weight Calculation for Performance Criteria (BSC Perspective).

Sr. No.	Performance criteria (BSC perspective)	1	2	3	4
1	Financial perspective	1	1/1.75	1.5	1.5
2	Customer perspective	1.75	1	1.5	2
3	Internal business perspective	1/1.5	1/1.5	1	1.5
4	Innovation and learning perspective	1/1.5	½	1/1.5	1

Sr. No.	Performance criteria (BSC perspective)	Weights
1	Financial perspective	0.257
2	Customer perspective	0.363
3	Internal business perspective	0.216
4	Innovation and learning perspective	0.163

Max. Eigen Value (λ max) = 4.002
Consistency Index = 0.006 and Consistency Ratio = 0.007.

TABLE 5.32 Tables of Pair-Wise Comparison and Weight Calculation for Performance Sub-criteria Under Financial Perspective.

Sr. No.	Sub-criteria under financial perspective	1	2	3	4	5	6
1	Rate of return on investment	1	1/2.7	1/3	1.25	1/1.75	1/1.25
2	Supplier's cost-saving initiatives	2.7	1	1/1.25	2.25	1.25	1.6
3	Information carrying cost	3	1.25	1	3.1	1.7	2.1
4	Cost per hour of operation	1/1.25	1/2.25	1/3.1	1	1/1.5	1/1.5
5	Inventory carrying cost	1.75	1/1.25	1/1.7	1.5	1	1.27
6	Manufacturing cost	1.25	1/1.6	1/ 2.1	1.5	1/ 1.27	1

Sr. No.	Performance sub-criteria under financial perspective	Weights
1	Rate of return on investment	0.099
2	Supplier's cost-saving initiatives	0.224
3	Information carrying cost	0.285
4	Cost per hour of operation	0.094
5	Inventory carrying cost	0.165
6	Manufacturing cost	0.133

Max. Eigen Value (λ max) = 6.005
Consistency Index = 0.001 and Consistency Ratio = 0.008.

TABLE 5.33 Tables of Pair-Wise Comparison and Weight Calculation for Performance Sub-Criteria Under Customer Perspective

Sr. No.	Sub-criteria under customer perspective	1	2	3	4	5	6
1	Quality of goods manufactured	1	1.5	3	1.25	2.6	2.24
2	Customer satisfaction	1/1.5	1	1.75	1/1.26	1.8	1.55
3	Customer query time	1/3	1/1.75	1	1/2	1/1.25	1/1.5
4	Delivery performance and lead time	1/1.25	1.26	2	1	1.83	1.5
5	Range of product and services	1/2.6	1/1.8	1.25	1/1.83	1	1/1.25
6	Effectiveness of distribution planning schedules	1/2.24	1/1.55	1.5	1/1.5	1.25	1

Sr. No.	Performance sub-criteria under customer perspective	Weights
1	Quality of goods delivered	0.280
2	Customer satisfaction	0.181
3	Customer query time	0.093
4	Delivery performance and lead time	0.207
5	Range of product and services	0.108
6	Effectiveness of distribution planning schedules	0.131

Max. Eigen Value (λ max) = 6.0002
Consistency Index = 0.0004 and Consistency Ratio = 0.0003.

TABLE 5.34 Tables of Pair-Wise Comparison and Weight Calculation for Performance Sub-Criteria Under Internal Business Process Perspective.

Sr. No.	Sub-criteria under internal business process perspective	1	2	3	4	5	6
1	Capacity utilization	1	1/1.5	1/1.95	1.25	1/1.25	1/ 2.75
2	Planned process cycle time	1.5	1	1/1.3	1.85	1.25	1/1.63
3	Total supply chain cycle time	1.95	1.3	1	2.1	1.5	1/1.25
4	Product development cycle time	1/1.25	1/1.85	1/2.1	1	1/1.5	1/3.1
5	Flexibility to meet customer needs	1.25	1/1.25	1/1.5	1.5	1	1/1.9
6	Extent of cooperation to improve quality	2.75	1.63	1.25	3.10	1.9	1

Sr. No.	Performance sub-criteria under internal business process perspective	Weights
1	Capacity utilization	0.113
2	Planned process cycle time	0.168
3	Total supply chain cycle time	0.210
4	product development cycle time	0.093
5	Flexibility to meet customer needs	0.139
6	Extent of cooperation to improve quality	0.276

Max. Eigen Value (λ max) = 6.008
Consistency Index = 0.0016 and Consistency Ratio = 0.0013.

TABLE 5.35 Tables of Pair-Wise Comparison and Weight Calculation for Performance Sub-Criteria under Innovation and Learning Perspective.

Sr. No.	Sub-criteria under innovation and learning perspective	1	2	3	4	5
1	Accuracy of forecasting	1	1.5	2.8	1.35	1.82
2	Buyer supplier partnership level	1/1.5	1	1.5	1/1.3	1.3
3	Employee satisfaction and skill orientation	1/2.8	1/1.5	1	1/2	1/1.3
4	Use of IT/KM technologies	1/1.35	1.3	2	1	1.5
5	Use of QE and QM techniques	1/1.82	1/1.3	1.30	1/1.5	1

Sr. No.	Performance sub-criteria innovation and learning perspective	Weights
1	Accuracy of forecasting	0.302
2	Buyer supplier partnership level	0.190
3	Employee satisfaction and skill orientation	0.117
4	Use of IT/KM technologies	0.236
5	Use of QE and QM techniques	0.155

Max. Eigen Value (λ max) = 5.13

Consistency Index = 0.032 and Consistency Ratio = 0.029.

Now, to find the impact of individual performance sub-criteria on the supply chain cyclic processes (alternatives in supply chain performance measurement framework).

Pair-wise comparison of alternatives with respect to performance criteria under financial perspective.

TABLE 5.36 Pair-Wise Comparison of Alternatives With Respect to Sub-Criteria (Rate ROI) for Analysis of the Impact on Overall Supply Chain Performance.

Sr. No.	Supply chain cyclic processes (alternatives)	1	2	3	Weights
1	Procurement	1	1/1.25	1/1.25	0.285
2	Production	1.25	1	1.25	0.384
3	Distribution	1.25	1/1.25	1	0.331

A: Sub-criteria: Rate of return on investment.

Max. Eigen Value (λ max) = 3.003

Consistency Index = 0.001 and Consistency Ratio = 0.0025.

TABLE 5.37 Pair-Wise Comparison of Alternatives With Respect to Sub-Criteria (SCSI) for Analysis of the Impact on Overall Supply Chain Performance.

Sr. No.	Supply chain cyclic processes (alternatives)	1	2	3	Weights
1	Procurement	1	2	1.5	0.462
2	Production	1/2	1	1/1.25	0.236
3	Distribution	1/1.5	1.25	1	0.302

B: Sub-criteria: Supplier's cost-saving initiatives.
Max. Eigen Value (λ max) = 3.001
Consistency Index = 0.0005 and Consistency Ratio = 0.0008.

TABLE 5.38 Pair-Wise Comparison of Alternatives With Respect to Sub-Criteria (Inf. C.C.) for Analysis of the Impact on Overall Supply Chain Performance.

Sr. No.	Supply chain cyclic processes (alternatives)	1	2	3	Weights
1	Procurement	1	1.5	1.1	0.378
2	Production	1/1.5	1	1.25	0.312
3	Distribution	1/1.1	1/1.25	1	0.307

C: Sub-criteria: Information carrying cost.
Max. Eigen Value (λ max) = 3.002
Consistency Index = 0.001 and Consistency Ratio = 0.0017.

TABLE 5.39 Pair-Wise Comparison of Alternatives With Respect to Sub-Criteria (Cost per Hour of Operation) for Analysis of the Impact on Overall Supply Chain Performance.

Sr. No.	Supply chain cyclic processes (alternatives)	1	2	3	Weights
1	Procurement	1	1/1.25	1.25	0.330
2	Production	1.25	1	1.25	0.385
3	Distribution	1/1.25	1/1.25	1	0.285

D: Sub-criteria: Cost per hour of operation.
Max. Eigen Value (λ max) = 3.004
Consistency Index = 0.002 and Consistency Ratio = 0.003.

TABLE 5.40 Pair-Wise Comparison of Alternatives With Respect to Sub-Criteria (Employee Satisfaction and Skill Orientation) for Analysis of the Impact on Overall Supply Chain Performance.

Sr. No.	Supply chain cyclic processes (alternatives)	1	2	3	Weights
1	Procurement	1	1.5	1.1	0.377
2	Production	1/1.5	1	1.5	0.331
3	Distribution	1/1.1	1/1.5	1	0.289

E: Sub-criteria: Inventory carrying cost.
Max. Eigen Value (λ max) = 3.003
Consistency Index = 0.001 and Consistency Ratio = 0.0025.

TABLE 5.41 Pair-Wise Comparison of Alternatives With Respect to Sub-Criteria (Manufacturing Cost) for Analysis of the Impact on Overall Supply Chain Performance.

Sr. No.	Supply chain cyclic processes (alternatives)	1	2	3	Weights
1	Procurement	1	1/1.25	1.5	0.347
2	Production	1.25	1	1.5	0.403
3	Distribution	1/1.5	1/1.5	1	0.249

F: Sub-criteria: Manufacturing cost.
Max. Eigen Value (λ max) = 3.006
Consistency Index = 0.003 and Consistency Ratio = 0.005.
Pair-wise comparison of alternatives with respect to performance criteria under customer perspective.

TABLE 5.42 Pair-Wise Comparison of Alternatives With Respect to Sub-Criteria (Quality of Goods Manufactured) for Analysis of the Impact on Overall Supply Chain Performance.

Sr. No.	Supply chain cyclic processes (alternatives)	1	2	3	Weights
1	Procurement	1	1.1	1.5	0.381
2	Production	1/1.1	1	1.5	0.369
3	Distribution	1/1.5	1/1.5	1	0.250

G: Sub-criteria: Quality of goods manufactured.
Max. Eigen Value (λ max) = 3.005
Consistency Index = 0.0025 and Consistency Ratio = 0.0043.

TABLE 5.43 Pair-Wise Comparison of Alternatives With Respect to Sub-Criteria (Deliver Performance and Lead Time) for Analysis of the Impact on Overall Supply Chain Performance.

Sr. No.	Supply chain cyclic processes (alternatives)	1	2	3	Weights
1	Procurement	1	1.1	1	0.332
2	Production	1/1.1	1	1.25	0.358
3	Distribution	1	1/1.25	1	0.309

H: Sub-criteria: Delivery performance and lead time.
Max. Eigen Value (λ max) = 3.0035
Consistency Index = 0.00175 and Consistency Ratio = 0.00301.

TABLE 5.44 Pair-Wise Comparison of Alternatives With Respect to Sub-Criteria (Cust. Satisfaction) for Analysis of the Impact on Overall Supply Chain Performance.

Sr. No.	Supply chain cyclic processes (alternatives)	1	2	3	Weights
1	Procurement	1	1.25	1	0.357
2	Production	1/1.25	1	1.5	0.352
3	Distribution	1	1/1.5	1	0.290

I: Sub-criteria: Customer satisfaction.
Max. Eigen Value (λ max) = 3.0022
Consistency Index = 0.0011 and Consistency Ratio = 0.00189.

TABLE 5.45 Pair-Wise Comparison of Alternatives With Respect to Sub-Criteria (EDPS) for Analysis of the Impact on Overall Supply Chain Performance.

Sr. No.	Supply chain cyclic processes (alternatives)	1	2	3	Weights
1	Procurement	1	1.5	1/1.25	0.347
2	Production	1/1.5	1	1/1.5	0.249
3	Distribution	1.25	1.5	1	0.403

J: Sub-criteria: Effectiveness of distribution planning schedules.
Max. Eigen Value (λ max) = 3.0018
Consistency Index = 0.0009 and Consistency Ratio = 0.0015.

TABLE 5.46 Pair-Wise Comparison of Alternatives With Respect to Sub-Criteria (CQT) for Analysis of the Impact on Overall Supply Chain Performance.

Sr. No.	Supply chain cyclic processes (alternatives)	1	2	3	Weights
1	Procurement	1	1.25	1/1.1	0.352
2	Production	1/1.25	1	1/1.25	0.286
3	Distribution	1.1	1.25	1	0.362

K: Sub-criteria: Customer query time.
Max. Eigen Value (λ max) = 3.0022
Consistency Index = 0.0011 and Consistency Ratio = 0.00189.

TABLE 5.47 Pair-Wise Comparison of Alternatives With Respect to Sub-Criteria (RP&S) for Analysis of the Impact on Overall Supply Chain Performance.

Sr. No.	Supply chain cyclic processes (alternatives)	1	2	3	Weights
1	Procurement	1	1.5	1.5	0.426
2	Production	1/1.5	1	1.5	0.325
3	Distribution	1/1.5	1/1.5	1	0.248

L: Sub-criteria: Range of product and services.
Max. Eigen Value (λ max) = 3.001
Consistency Index = 0.0005 and Consistency Ratio = 0.00086.
Pair-wise comparison of alternatives with respect to performance criteria under internal business process perspective.

TABLE 5.48 Pair-Wise Comparison of Alternatives With Respect to Sub-Criteria (Flex. to Meet Particular Customer Needs) for Analysis of the Impact on Overall Supply Chain Performance.

Sr. No.	Supply chain cyclic processes (alternatives)	1	2	3	Weights
1	Procurement	1	1/1.1	1.5	0.363
2	Production	1.1	1	1.5	0.387
3	Distribution	1/1.5	1/1.5	1	0.250

M: Sub-criteria: Flexibility to meet particular customer needs.
Max. Eigen Value (λ max) = 3.002
Consistency Index = 0.001 and Consistency Ratio = 0.0017.

TABLE 5.49 Pair-Wise Comparison of Alternatives With Respect to Sub-Criteria (TSCCT) for Analysis of the Impact on Overall Supply Chain Performance.

Sr. No.	Supply chain cyclic processes (alternatives)	1	2	3	Weights
1	Procurement	1	1.5	1.27	0.408
2	Production	1/1.5	1	1.26	0.311
3	Distribution	1/1.27	1/1.26	1	0.281

N: Sub-criteria: Total supply chain cycle time.
Max. Eigen Value (λ max) = 3.001
Consistency Index = 0.0005 and Consistency Ratio = 0.00086.

TABLE 5.50 Pair-Wise Comparison of Alternatives With Respect to Sub-Criteria (PDCT) for Analysis of the Impact on Overall Supply Chain Performance.

Sr. No.	Supply chain cyclic processes (alternatives)	1	2	3	Weights
1	Procurement	1	1/1.25	1.5	0.350
2	Production	1.25	1	1.25	0.382
3	Distribution	1/1.5	1/1.25	1	0.267

O: Sub-criteria: Product development cycle time.
Max. Eigen Value (λ max) = 3.0013
Consistency Index = 0.00065 and Consistency Ratio = 0.0011.

TABLE 5.51 Pair-Wise Comparison of Alternatives With Respect to Sub-Criteria (CU) for Analysis of the Impact on Overall Supply Chain Performance.

Sr. No.	Supply chain cyclic processes (alternatives)	1	2	3	Weights
1	Procurement	1	1/1.5	1/1.25	0.266
2	Production	1.5	1	1.25	0.405
3	Distribution	1.25	1/1.25	1	0.329

P: Sub-criteria: Capacity utilization.
Max. Eigen Value (λ max) = 3.0004
Consistency Index = 0.0002 and Consistency Ratio = 0.00034.

TABLE 5.52 Pair-Wise Comparison of Alternatives With Respect to Sub-Criteria (PPCT) for Analysis of the Impact on Overall Supply Chain Performance.

Sr. No.	Supply chain cyclic processes (alternatives)	1	2	3	Weights
1	Procurement	1	1.25	1.25	0.382
2	Production	1/1.25	1	1.5	0.350
3	Distribution	1/1.25	1/1.5	1	0.267

Q: Sub-criteria: Planned process cycle time.
Max. Eigen Value (λ max) = 3.001
Consistency Index = 0.0005 and Consistency Ratio = 0.00086.

TABLE 5.53 Pair-Wise Comparison of Alternatives With Respect to Sub-Criteria (Employee Satisfaction and Skill Orientation) for Analysis of the Impact on Overall Supply Chain Performance.

Sr. No.	Supply chain cyclic processes (alternatives)	1	2	3	Weights
1	Procurement	1	1.24	1.5	0.403
2	Production	1/1.24	1	1.5	0.350
3	Distribution	1/1.5	1/1.5	1	0.247

R: Sub-criteria: Extent of cooperation to improve quality.
Max. Eigen Value (λ max) = 3.0006
Consistency Index = 0.0003 and Consistency Ratio = 0.00051.
Pair-wise comparison of alternatives with respect to performance criteria under innovation and learning perspective.

TABLE 5.54 Pair-Wise Comparison of Alternatives With Respect to Sub-Criteria (AoF) for Analysis of the Impact on Overall Supply Chain Performance.

Sr. No.	Supply chain cyclic processes (alternatives)	1	2	3	Weights
1	Procurement	1	2	1.25	0.439
2	Production	1/2	1	1/1.25	0.238
3	Distribution	1/1.25	1.25	1	0.323

S: Sub-criteria: Accuracy of forecasting.
Max. Eigen Value (λ max) = 3.002
Consistency Index = 0.001 and Consistency Ratio=0.0017.

TABLE 5.55 Pair-Wise Comparison of Alternatives With Respect to Sub-Criteria (BSPL) for Analysis of the Impact on Overall Supply Chain Performance.

Sr. No.	Supply chain cyclic processes (alternatives)	1	2	3	Weights
1	Procurement	1	2	1.1	0.419
2	Production	1/2	1	1/1.5	0.225
3	Distribution	1/1.1	1.5	1	0.357

T: Sub-criteria: Buyer supplier partnership level.
Max. Eigen Value (λ max) = 3.005
Consistency Index = 0.0025 and Consistency Ratio=0.0043.

TABLE 5.56 Pair-Wise Comparison of Alternatives With Respect to Sub-Criteria (IT and KM) for Analysis of the Impact on Overall Supply Chain Performance.

Sr. No.	Supply chain cyclic processes (alternatives)	1	2	3	Weights
1	Procurement	1	1.7	1.3	0.425
2	Production	1/1.7	1	1.5	0.313
3	Distribution	1/1.3	1/1.5	1	0.262

U: Sub-criteria: Use of IT and KM concepts
Max. Eigen Value (λ max) = 3.003
Consistency Index = 0.0015 and Consistency Ratio= 0.0025.

TABLE 5.57 Pair-Wise Comparison of Alternatives With Respect to Sub-Criteria (QE and QM) for Analysis of the Impact on Overall Supply Chain Performance.

Sr. No.	Supply chain cyclic processes (alternatives)	1	2	3	Weights
1	Procurement	1	1/1.25	1	0.306
2	Production	1.25	1	1.5	0.407
3	Distribution	1	1/1.5	1	0.287

V: Sub-criteria: Use of QE and QM techniques.
Max. Eigen Value (λ max) = 3.001
Consistency Index = 0.0005 and Consistency Ratio = 0.00086.

TABLE 5.58 Pair-Wise Comparison of Alternatives With Respect to Sub-Criteria (Emp. S. and Skill O.) for Analysis of the Impact on Overall Supply Chain Performance.

Sr. No.	Supply chain cyclic processes (alternatives)	1	2	3	Weights
1	Procurement	1	1.2	1.1	0.363
2	Production	1/1.2	1	1.5	0.356
3	Distribution	1/1.1	1/1.5	1	0.281

W: Sub-criteria: Employee satisfaction and skill orientation.
Max. Eigen Value (λ max) = 3.0011
Consistency Index = 0.00055 and Consistency Ratio = 0.00094.

TABLE 5.59 Importance of Supply Chain Cyclic Processes With Respect to Overall Supply Chain Performance Measurement.

Sr. No.	Supply chain cyclic processes (alternatives)	Final Weight
1	Procurement	0.377
2	Production	0.331
3	Distribution	0.292

TABLE 5.60 Appendix C: Comparative Priorities of Supply Chain Cyclic Processes With Respect to Various Performance Measures (Sub-Criteria).

	Financial Perspective						Customer Perspective					
Level of Sub-criteria	St	Tt	Op	Op	Op	Op	Op	St	S, T, O	Tt	St	St
Supply Chain Cyclic Processes	A	B	C	D	E	F	G	H	I	J	K	L
Procurement	0.285	0.462	0.378	0.33	0.377	0.347	0.381	0.332	0.357	0.347	0.352	0.426
Production	0.384	0.236	0.312	0.385	0.331	0.403	0.369	0.358	0.352	0.249	0.296	0.325
Distribution	0.331	0.302	0.307	0.285	0.289	0.249	0.25	0.309	0.29	0.403	0.362	0.248
Max. Eigen Value (λ max)	3.003	3.001	3.002	3.004	3.003	3.006	3.005	3.0035	3.0022	3.0018	3.0022	3.001
Consistency Index	0.001	0.0005	0.001	0.002	0.001	0.003	0.0025	0.00175	0.0011	0.0009	0.0011	0.0005
Consistency Ratio	0.0025	0.0008	0.0017	0.003	0.0025	0.005	0.0043	0.00301	0.00189	0.0015	0.00189	0.00086

	Internal Business Process Perspective						Innovation & Learning Perspective				
Level of Sub-criteria	St	St	Tt	Op	Tt	S, T, O	St, Tt	St	S, T, O	S, T, O	S, T, O
Supply Chain Cyclic Processes	M	N	O	P	Q	R	S	T	U	V	W
Procurement	0.363	0.408	0.35	0.265	0.382	0.403	0.439	0.419	0.425	0.306	0.363
Production	0.387	0.311	0.382	0.405	0.35	0.35	0.238	0.225	0.313	0.407	0.356
Distribution	0.25	0.281	0.267	0.329	0.267	0.247	0.323	0.357	0.262	0.287	0.281
Max. Eigen Value (λ max)	3.002	3.001	3.0013	3.0034	3.001	3.0006	3.002	3.005	3.003	3.001	3.0011
Consistency Index	0.001	0.0005	0.00065	0.0002	0.0005	0.0003	0.001	0.0025	0.0015	0.0005	0.00055
Consistency Ratio	0.0017	0.00086	0.0011	0.00034	0.00086	0.00051	0.0017	0.0043	0.0025	0.00086	0.00094

5.5.3 *CASE COMPANY NO. 3*

We first compare the performance criteria for estimating their relative weight.

TABLE 5.61 Tables of Pair-Wise Comparison and Weight Calculation for Performance Criteria (BSC Perspective).

Sr. No.	Performance criteria (BSC perspective)	1	2	3	4
1	Financial perspective	1	1/1.3	1/2	1.3
2	Customer perspective	1.3	1	1/1.4	1.5
3	Internal business perspective	2	1.4	1	2
4	Innovation and learning perspective	1/1.3	1/1.5	1/2	1

Sr. No.	Performance criteria (BSC perspective)	Weights
1	Financial perspective	0.201
2	Customer perspective	0.262
3	Internal business perspective	0.366
4	Innovation and learning perspective	0.170

Max. Eigen Value (λ max) = 4.001
Consistency Index = 0.00033 and Consistency Ratio = 0.00036.

TABLE 5.62 Tables of Pair-Wise Comparison and Weight Calculation for Performance Sub-Criteria Under Financial Perspective.

Sr. No.	Sub-criteria under financial perspective	1	2	3	4	5	6
1	Rate of return on investment	1	1/1.36	1/ 2.1	1.5	1/2.15	1.25
2	Supplier's cost-saving initiatives	1.36	1	1/1.25	1.75	1/1.75	1.5
3	Information carrying cost	2.1	1.25	1	2	1/1.25	1.8
4	Cost per hour of operation	1/1.5	1/1.75	1/2	1	1/3	1/1.25
5	Inventory carrying cost	2.15	1.75	1.25	3	1	2
6	Manufacturing cost	1/1.25	1/1.5	1/1.8	1.25	1/2	1

Sr. No.	Performance sub-criteria under financial perspective	Weights
1	Rate of return on investment	0.128
2	Supplier's cost-saving initiatives	0.169
3	Information carrying cost	0.219
4	Cost per hour of operation	0.094
5	Inventory carrying cost	0.273
6	Manufacturing cost	0.117

Max. Eigen Value (λ max) = 6.001
Consistency Index = 0.0002 and Consistency Ratio = 0.00016.

TABLE 5.63 Tables of Pair-Wise Comparison and Weight Calculation for Performance Sub-Criteria Under Customer Perspective.

Sr. No.	Sub-criteria under customer perspective	1	2	3	4	5	6
1	Quality of goods manufactured	1	1.5	2.16	1.8	3.1	1.9
2	Customer satisfaction	1/1.5	1	1.82	1.25	2	1.5
3	Customer query time	1/ 2.16	1/1.82	1	1/1.5	1.3	1/1.3
4	Delivery performance and lead	1/1.8	1/1.25	1.5	1	1.75	1.25
5	Range of product and services	1/3.1	1/ 2	1/1.3	1/1.75	1	1/1.5
6	Effectiveness of distribution planning schedules	1/1.9	1/1.5	1/1.3	1/1.25	1.5	1

Sr. No.	Performance sub-criteria under customer perspective	Weights
1	Quality of goods delivered	0.283
2	Customer satisfaction	0.202
3	Customer query time	0.116
4	Delivery performance and lead	0.166
5	Range of product and services	0.093
6	Effectiveness of distribution planning schedules	0.140

Max. Eigen Value (λ max) = 6.002
Consistency Index = 0.0004 and Consistency Ratio = 0.00032.

TABLE 5.64 Tables of Pair-Wise Comparison and Weight Calculation for Performance Sub-Criteria Under Internal Business Process Perspective.

Sr. No.	Sub-criteria under internal business process perspective	1	2	3	4	5	6
1	Capacity utilization	1	1/1.5	1/1.3	1.5	1.3	1/3
2	Planned process cycle time	1.5	1	1.3	2	1.8	1/1.25
3	Total supply chain cycle time	1.3	1/1.3	1	1.9	1.5	1/1.5
4	Product development cycle time	1/1.5	1/ 2	1/1.9	1	1/1.15	1/3
5	Flexibility to meet customer needs	1/1.3	1/1.8	1/1.5	1.15	1	1/ 2.5
6	Extent of cooperation to improve quality	3	1.25	1.5	3	2.5	1

Sr. No.	Performance sub-criteria under internal business process perspective	Weights
1	Capacity utilization	0.130
2	Planned process cycle time	0.205
3	Total supply chain cycle time	0.172
4	Product development cycle time	0.093
5	Flexibility to meet customer needs	0.108
6	Extent of cooperation to improve quality	0.292

Max. Eigen Value (λ max) = 6.0031
Consistency Index = 0.00062 and Consistency Ratio = 0.0005.

TABLE 5.65 Tables of Pair-Wise Comparison and Weight Calculation for Performance Sub-Criteria Under Innovation and Learning Perspective.

Sr. No.	Sub-criteria under innovation and learning perspective	1	2	3	4	5
1	Accuracy of forecasting	1	1/1.5	1.5	2	1.5
2	Buyer supplier partnership level	1.5	1	2.5	2	1.8
3	Employee satisfaction and skill orientation	1/1.5	1/ 2.5	1	1/1.35	1/1.8
4	Use of IT and KM technologies	1/ 2	1/ 2	1.35	1	1/1.25
5	Use of QE and QM techniques	1/1.5	1/ 1.8	1.8	1.25	1

Sr. No.	Performance sub-criteria innovation and learning perspective	Weights
1	Accuracy of forecasting	0.237
2	Buyer supplier partnership level	0.316
3	Employee satisfaction and skill orientation	0.122
4	Use of IT and KM technologies	0.145
5	Use of QE and QM techniques	0.180

Max. Eigen Value (λ max) = 5.001
Consistency Index = 0.0025 and Consistency Ratio = 0.0022.
Now, to find the impact of individual performance sub-criteria on the supply chain cyclic processes (alternatives in supply chain performance measurement framework).
Pair-wise comparison of alternatives with respect to performance criteria under financial perspective.

TABLE 5.66 Pair-Wise Comparison of Alternatives With Respect to Sub-Criteria (Rate of ROI) for Analysis of the Impact on Overall Supply Chain Performance.

Sr. No.	Supply chain cyclic processes (alternatives)	1	2	3	Weights
1	Procurement	1	1/1.6	1/1.5	0.239
2	Production	1.6	1	2	0.471
3	Distribution	1.5	1/ 2	1	0.290

A: Sub-criteria: Rate of return on investment.
Max. Eigen Value (λ max) = 3.008
Consistency Index = 0.004 and Consistency Ratio = 0.006.

TABLE 5.67 Pair-Wise Comparison of Alternatives With Respect to Sub-Criteria (SCSI) for Analysis of the Impact on Overall Supply Chain Performance.

Sr. No.	Supply chain cyclic processes (alternatives)	1	2	3	Weights
1	Procurement	1	1.25	1	0.361
2	Production	1/1.25	1	1/1.3	0.281
3	Distribution	1	1.3	1	0.358

B: Sub-criteria: Supplier's cost-saving initiatives.
Max. Eigen Value (λ max) = 3.0008
Consistency Index = 0.0004 and Consistency Ratio = 0.0006.

TABLE 5.68 Pair-Wise Comparison of Alternatives with Respect to Sub-Criteria (Info. C. Cost) for Analysis of the Impact on Overall Supply Chain Performance.

Sr. No.	Supply chain cyclic processes (alternatives)	1	2	3	Weights
1	Procurement	1	1/1.4	1.25	0.309
2	Production	1.4	1	2	0.454
3	Distribution	1/1.25	1/ 2	1	0.237

C: Sub-criteria: Information carrying cost.
Max. Eigen Value (λ max) = 3.004
Consistency Index = 0.002 and Consistency Ratio = 0.0034.

TABLE 5.69 Pair-Wise Comparison of Alternatives With Respect to Sub-Criteria (Cost/h of Operation) for Analysis of the Impact on Overall Supply Chain Performance.

Sr. No.	Supply chain cyclic processes (alternatives)	1	2	3	Weights
1	Procurement	1	1/1.8	1/1.2	0.243
2	Production	1.8	1	2.25	0.501
3	Distribution	1.2	1/ 2.25	1	0.255

D: Sub-criteria: Cost per hour of operation.
Max. Eigen Value (λ max) = 3.004
Consistency Index = 0.002 and Consistency Ratio = 0.0034.

TABLE 5.70 Pair-Wise Comparison of Alternatives With Respect to Sub-Criteria (Inv. C. Cost) for Analysis of the Impact on Overall Supply Chain Performance.

Sr. No.	Supply chain cyclic processes (alternatives)	1	2	3	Weights
1	Procurement	1	1/1.75	1/1.5	0.228
2	Production	1.75	1	2.3	0.499
3	Distribution	1.5	1/ 2.3	1	0.273

E: Sub-criteria: Inventory carrying cost.
Max. Eigen Value (λ max) = 3.1
Consistency Index = 0.05 and Consistency Ratio = 0.086.

TABLE 5.71 Pair-Wise Comparison of Alternatives With Respect to Sub-Criteria (Manufacturing Cost) for Analysis of the Impact on Overall Supply Chain Performance.

Sr. No.	Supply chain cyclic processes (alternatives)	1	2	3	Weights
1	Procurement	1	1/ 2.5	1	0.225
2	Production	2.5	1	2.25	0.542
3	Distribution	1	1/ 2.25	1	0.233

F: Sub-criteria: Manufacturing cost.
Max. Eigen Value (λ max) = 3.002
Consistency Index = 0.001 and Consistency Ratio= 0.0017.
Pair-wise comparison of alternatives with respect to performance criteria under customer perspective.

TABLE 5.72 Pair-Wise Comparison of Alternatives With Respect to Sub-Criteria (QGM) for Analysis of the Impact on Overall Supply Chain Performance.

Sr. No.	Supply chain cyclic processes (alternatives)	1	2	3	Weights
1	Procurement	1	1/1.5	1/1.25	0.263
2	Production	1.5	1	1.7	0.444
3	Distribution	1.25	1/1.7	1	0.293

G: Sub-criteria: Quality of goods manufactured.
Max. Eigen Value (λ max) = 3.006
Consistency Index = 0.003 and Consistency Ratio = 0.0051.

TABLE 5.73 Pair-Wise Comparison of Alternatives With Respect to Sub-Criteria (DP and LT) for Analysis of the Impact on Overall Supply Chain Performance.

Sr. No.	Supply chain cyclic processes (alternatives)	1	2	3	Weights
1	Procurement	1	1/1.8	1/1.4	0.233
2	Production	1.8	1	2	0.486
3	Distribution	1.4	1/ 2	1	0.281

H: Sub-criteria: Delivery performance and lead time.
Max. Eigen Value (λ max) = 3.004
Consistency Index = 0.002 and Consistency Ratio = 0.0034.

TABLE 5.74 Pair-Wise Comparison of Alternatives With Respect to Sub-Criteria (Cust. Satisfaction) for Analysis of the Impact on Overall Supply Chain Performance.

Sr. No.	Supply chain cyclic processes (alternatives)	1	2	3	Weights
1	Procurement	1	1/1.4	1.1	0.303
2	Production	1.4	1	1.6	0.428
3	Distribution	1/1.1	1/1.6	1	0.269

I: Sub-criteria: Customer satisfaction.
Max. Eigen Value (λ max) = 3.008
Consistency Index = 0.004 and Consistency Ratio = 0.0068.

TABLE 5.75 Pair-Wise Comparisons of Alternatives With Respect to Sub-Criteria (EDPS) for Analysis of the Impact on Overall Supply Chain Performance.

Sr. No.	Supply chain cyclic processes (alternatives)	1	2	3	Weights
1	Procurement	1	1.5	1/1.5	0.322
2	Production	1/1.5	1	1/1.75	0.234
3	Distribution	1.5	1.75	1	0.444

J: Sub-criteria: Effectiveness of distribution planning schedules.
Max. Eigen Value (λ max) = 3.0132
Consistency Index = 0.0066 and Consistency Ratio = 0.011.

TABLE 5.76 Pair-Wise Comparison of Alternatives With Respect to Sub-Criteria (CQT) for Analysis of the Impact on Overall Supply Chain Performance.

Sr. No.	Supply chain cyclic processes (alternatives)	1	2	3	Weights
1	Procurement	1	1.25	1/1.28	0.329
2	Production	1/1.25	1	1/1.2	0.289
3	Distribution	1.28	1.2	1	0.382

K: Sub-criteria: Customer query time.
Max. Eigen Value (λ max) = 3.016
Consistency Index = 0.008 and Consistency Ratio = 0.013.

TABLE 5.77 Pair-Wise Comparison of Alternatives With Respect to Sub-Criteria (RP&S) for Analysis of the Impact on Overall Supply Chain Performance.

Sr. No.	Supply chain cyclic processes (alternatives)	1	2	3	Weights
1	Procurement	1	1/ 2	1.5	0.292
2	Production	2	1	1.5	0.463
3	Distribution	1/1.5	1/1.5	1	0.245

L: Sub-criteria: Range of product and services.
Max. Eigen Value (λ max) = 3.1
Consistency Index = 0.05 and Consistency Ratio = 0.086
Pair-wise comparison of alternatives with respect to performance criteria under internal business process perspective.

TABLE 5.78 Pair-Wise Comparison of Alternatives With Respect to Sub-Criteria (Flex. To Meet Cust. Needs) for Analysis of the Impact on Overall Supply Chain Performance.

Sr. No.	Supply Chain Cyclic Processes (Alternatives)	1	2	3	Weights
1	Procurement	1	1/1.8	1.3	0.288
2	Production	1.8	1	1.7	0.466
3	Distribution	1/1.3	1/1.7	1	0.246

M: Sub-criteria: Flexibility to Meet Particular Customer Needs.
Max. Eigen Value (λ max) = 3.02
Consistency Index = 0.01 and Consistency Ratio = 0.0174.

TABLE 5.79 Pair-Wise Comparison of Alternatives With Respect to Sub-Criteria (QGM) for Analysis of the Impact on Overall Supply Chain Performance.

Sr. No.	Supply chain cyclic processes (alternatives)	1	2	3	Weights
1	Procurement	1	1.25	1.1	0.368
2	Production	1/1.25	1	1.5	0.351
3	Distribution	1/1.1	1/1.5	1	0.281

N: Sub-criteria: Total supply chain cycle time.
Max. Eigen Value (λ max) = 3.06
Consistency Index = 0.03 and Consistency Ratio = 0.051.

TABLE 5.80 Pair-Wise Comparison of Alternatives With Respect to Sub-Criteria (PDCT) for Analysis of the Impact on Overall Supply Chain Performance.

Sr. No.	Supply chain cyclic processes (alternatives)	1	2	3	Weights
1	Procurement	1	1/1.5	1.1	0.293
2	Production	1.5	1	1.7	0.444
3	Distribution	1/1.1	1/1.7	1	0263

O: Sub-criteria: Product development cycle time.
Max. Eigen Value (λ max) = 3.0018
Consistency Index = 0.0009 and Consistency Ratio = 0.00155.

TABLE 5.81 Pair-Wise Comparison of Alternatives With Respect to Sub-Criteria (CU) for Analysis of the Impact on Overall Supply Chain Performance

Sr. No.	Supply chain cyclic processes (alternatives)	1	2	3	Weights
1	Procurement	1	1/ 2.5	1/ 2	0.181
2	Production	2.5	1	1.5	0.480
3	Distribution	2	1/1.5	1	0.340

P: Sub-criteria: Capacity utilization.
Max. Eigen Value (λ max) = 3.006
Consistency Index = 0.003 and Consistency Ratio = 0.0051.

TABLE 5.82 Pair-Wise Comparison of Alternatives With Respect to Sub-Criteria (PPCT) for Analysis of the Impact on Overall Supply Chain Performance

Sr. No.	Supply chain cyclic processes (Alternatives)	1	2	3	Weights
1	Procurement	1	1/1.1	1.25	0.344
2	Production	1.1	1	1.5	0.389
3	Distribution	1/1.25	1/1.5	1	0.267

Q: Sub-criteria: Planned process cycle time.
Max. Eigen Value (λ max) = 3.0016
Consistency Index = 0.0008 and Consistency Ratio = 0.0013.

TABLE 5.83 Pair-Wise Comparison of Alternatives With Respect to Sub-Criteria (Ext. of Cooperation to Improve Quality) for Analysis of the Impact on Overall Supply Chain Performance.

Sr. No.	Supply chain cyclic processes (alternatives)	1	2	3	Weights
1	Procurement	1	1/1.1	1.25	0.325
2	Production	1.1	1	1.5	0.495
3	Distribution	1/1.25	1/1.5	1	0.180

R: Sub-criteria: Extent of cooperation to improve quality
Max. Eigen Value (λ max) = 3.016
Consistency Index = 0.008 and Consistency Ratio = 0.013.
Pair-wise comparison of alternatives with respect to performance criteria under innovation and learning perspective.

TABLE 5.84 Pair-Wise Comparison of Alternatives With Respect to Sub-Criteria (AoF) for Analysis of the Impact on Overall Supply Chain Performance

Sr. No.	Supply chain cyclic processes (alternatives)	1	2	3	Weights
1	Procurement	1	1.8	1.1	0.407
2	Production	1/1.8	1	1/1.5	0.233
3	Distribution	1/1.1	1.5	1	0.360

S: Sub-criteria: Accuracy of forecasting.
Max. Eigen Value (λ max) = 3.014
Consistency Index = 0.007 and Consistency Ratio = 0.012.

TABLE 5.85 Pair-Wise Comparison of Alternatives With Respect to Sub-Criteria (BSPL) for Analysis of the Impact on Overall Supply Chain Performance.

Sr. No.	Supply chain cyclic processes (alternatives)	1	2	3	Weights
1	Procurement	1	2.5	1	0.428
2	Production	1/ 2.5	1	1/1.7	0.194
3	Distribution	1	1.7	1	0.377

T: Sub-criteria: Buyer supplier partnership level.
Max. Eigen Value (λ max) = 3.04
Consistency Index = 0.02 and Consistency Ratio = 0.034.

TABLE 5.86 Pair-Wise Comparison of Alternatives With Respect to Sub-Criteria (IT and KM) for Analysis of the Impact on Overall Supply Chain Performance.

Sr. No.	Supply chain cyclic processes (alternatives)	1	2	3	Weights
1	Procurement	1	1/1.5	1.25	0.304
2	Production	1.5	1	1.75	0.446
3	Distribution	1/1.25	1/1.75	1	0.249

U: Sub-criteria: Use of IT and KM concepts.
Max. Eigen Value (λ max) = 3.001
Consistency Index = 0.0005 and Consistency Ratio = 0.00086.

TABLE 5.87 Pair-Wise Comparison of Alternatives With Respect to Sub-Criteria (QE and QM) for Analysis of the Impact on Overall Supply Chain Performance.

Sr. No.	Supply chain cyclic processes (alternatives)	1	2	3	Weights
1	Procurement	1	1/1.75	1/1.1	0.260
2	Production	1.75	1	1.5	0.448
3	Distribution	1.1	1/1.5	1	0.292

V: Sub-criteria: Use of QE and QM techniques.
Max. Eigen Value (λ max) = 3.0006
Consistency Index = 0.0003 and Consistency Ratio = 0.00051.

TABLE 5.88 Pair-Wise Comparison of Alternatives With Respect to Sub-Criteria (ESSO) for Analysis of the Impact on Overall Supply Chain Performance.

Sr. No.	Supply chain cyclic processes (alternatives)	1	2	3	Weights
1	Procurement	1	1/1.25	1/1.25	0.285
2	Production	1.25	1	1.25	0.382
3	Distribution	1.25	1/1.25	1	0.333

W: Sub-criteria: Employee satisfaction and skill orientation.
Max. Eigen Value (λ max) = 3.008
Consistency Index = 0.004 and Consistency Ratio = 0.0068

TABLE 5.89 Importance of Supply Chain Cyclic Processes With Respect to Overall Supply Chain Performance Measurement.

Sr. No.	Supply chain cyclic processes (alternatives)	Final weight
1	Procurement	0.301
2	Production	0.409
3	Distribution	0.290

TABLE 5.90 Appendix D: Comparative Priorities of Supply Chain Cyclic Processes With Respect to Various Performance Measures (Sub-Criteria).

	Financial Perspective						Customer Perspective					
Level of Sub-criteria	St	Tt	Op	Op	Op	Op	Op	St	S, T, O	Tt	St	St
Supply Chain Cyclic Processes	A	B	C	D	E	F	G	H	I	J	K	L
Procurement	0.239	0.361	0.309	0.243	0.228	0.225	0.263	0.233	0.303	0.322	0.329	0.292
Production	0.471	0.281	0.454	0.501	0.499	0.542	0.444	0.486	0.428	0.234	0.289	0.463
Distribution	0.29	0.358	0.237	0.255	0.273	0.233	0.293	0.281	0.269	0.444	0.382	0.245
Max. Eigen Value (λ max)	3.008	3.0008	3.004	3.004	3.1	3.002	3.006	3.004	3.008	3.0132	3.016	3.1
Consistency Index	0.004	0.0004	0.002	0.002	0.05	0.001	0.003	0.002	0.004	0.0066	0.008	0.05
Consistency Ratio	0.006	0.0006	0.0034	0.0034	0.086	0.0017	0.005	0.0034	0.0068	0.011	0.013	0.086

	Internal Business Process Perspective						Innovation & Learning Perspective				
Level of Sub-criteria	St	St	Tt	Op	Tt	S, T, O	St, Tt	St	S, T, O	S, T, O	S, T, O
Supply Chain Cyclic Processes	M	N	O	P	Q	R	S	T	U	V	W
Procurement	0.288	0.368	0.293	0.181	0.344	0.344	0.407	0.428	0.304	0.26	0.285
Production	0.466	0.351	0.444	0.48	0.389	0.389	0.233	0.194	0.446	0.448	0.382
Distribution	0.246	0.281	263	0.34	0.267	0.267	0.36	0.377	0.249	0.292	0.333
Max. Eigen Value (λ max)	3.02	3.06	3.0018	3.006	3.0016	3.016	3.014	3.04	3.001	3.0006	3.008
Consistency Index	0.01	0.03	0.0009	0.003	0.0008	0.008	0.007	0.02	0.0005	0.0003	0.004
Consistency Ratio	0.017	0.051	0.0016	0.0051	0.0013	0.013	0.012	0.034	0.0009	0.00051	0.0068

5.5.4 *CASE COMPANY NO. 4*

We first compare the performance criteria for estimating their relative weight.

TABLE 5.91 Tables of Pair-Wise Comparison and Weight Calculation for Performance Criteria (BSC Perspective).

Sr. No.	Performance criteria (BSC perspective)	1	2	3	4
1	Financial perspective	1	1.4	1/1.7	1.5
2	Customer perspective	1/1.4	1	1/ 2	1.25
3	Internal business perspective	1.7	2	1	2.5
4	Innovation and learning perspective	1/1.5	1/ 1.25	1/2.5	1

Sr. No.	Performance criteria (BSC perspective)	Weights
1	Financial perspective	0.248
2	Customer perspective	0.192
3	Internal business perspective	0.401
4	Innovation and learning perspective	0.160

Max. Eigen Value (λ max) = 4.003
Consistency Index = 0.001 and Consistency Ratio = 0.0011.

TABLE 5.92 Tables of Pair-Wise Comparison and Weight Calculation for Performance Sub-Criteria Under Financial Perspective

Sr. No.	Sub-criteria under financial perspective	1	2	3	4	5	6
1	Rate of return on investment	1	1/1.5	1/2	1.5	1 /2	1.25
2	Supplier's cost-saving initiatives	1.5	1	1/1.3	1.5	1/1.5	1.6
3	Information carrying cost	2	1.3	1	2.5	1/ 1.25	2
4	Cost per hour of operation	1/1.5	1/1.5	1/ 2.5	1	1/3	1/ 1.2
5	Inventory carrying cost	2	1.5	1.25	3	1	2.5
6	Manufacturing cost	1/1.25	1/1.6	1 /2	1.2	1/ 2.5	1

Sr. No.	Performance sub-criteria under financial perspective	Weights
1	Rate of return on investment	0.127
2	Supplier's cost-saving initiatives	0.171
3	Information carrying cost	0.228
4	Cost per hour of operation	0.094
5	Inventory carrying cost	0.270
6	Manufacturing cost	0.108

Max. Eigen Value (λ max) = 6.012
Consistency Index = 0.0024 and Consistency Ratio = 0.019.

TABLE 5.93 Tables of Pair-Wise Comparison and Weight Calculation for Performance Sub-Criteria Under Customer Perspective.

Sr. No.	Sub-criteria under customer perspective	1	2	3	4	5	6
1	Quality of goods manufactured	1	1.25	2.5	1.75	3	2
2	Customer satisfaction	1/ 1.25	1	2.55	1.25	2	1.5
3	Customer query time	1/ 2.5	1/ 2.55	1	1/1.5	1.1	1/ 1.25
4	Delivery performance and lead time	1/ 1.75	1/ 1.25	1.5	1	1.75	1.2
5	Range of product and services	1/3	1/ 2	1/1.1	1/ 1.75	1	1/1.5
6	Effectiveness of distribution planning schedules	1/ 2	1/ 1.5	1.25	1/1.2	1.5	1

Sr. No.	Performance sub-criteria under customer perspective	Weights
1	Quality of goods manufactured	0.277
2	Customer satisfaction	0.218
3	Customer query time	0.104
4	Delivery performance and lead time	0.165
5	Range of product and services	0.096
6	Effectiveness of distribution planning schedules	0.139

Max. Eigen Value (λ max) = 6.013
Consistency Index = 0.0026 and Consistency Ratio = 0.0020.

TABLE 5.94 Tables of Pair-Wise Comparison and Weight Calculation for Performance Sub-Criteria under Internal Business Process Perspective.

Sr. No.	Sub-criteria under internal business process perspective	1	2	3	4	5	6
1	Capacity utilization	1	1/1.5	1/1.25	1.5	1.25	1/ 2
2	Planned process cycle time	1.5	1	1.3	2	1.75	1/1.3
3	Total supply chain cycle time	1.25	1/1.3	1	1.75	1.5	1/1.5
4	Product development cycle time	1/1.5	1/ 2	1/1.75	1	1/1.25	1/3
5	Flexibility to meet particular customer needs	1/1.25	1/1.75	1/1.5	1.25	1	1/ 2.5
6	Extent of cooperation to improve quality	2	1.3	1.5	3	2.5	1

Sr. No.	Performance sub-criteria under internal business process perspective	Weights
1	Capacity utilization	0.139
2	Planned process cycle time	0.206
3	Total supply chain cycle time	0.170
4	Product development cycle time	0.095
5	Flexibility to meet particular customer needs	0.114
6	Extent of cooperation to improve quality	0.275

Max. Eigen Value (λ max) = 6.0034
Consistency Index =0.00068 and Consistency Ratio = 0.0054.

TABLE 5.95 Tables of Pair-Wise Comparison and Weight Calculation for Performance Sub-Criteria under Innovation and Learning Perspective.

Sr. No.	Sub-criteria under innovation and learning perspective	1	2	3	4	5
1	Accuracy of forecasting	1	1/ 1.3	2	1.7	1.3
2	Buyer supplier partnership level	1.3	1	3	2.25	1.6
3	Employee satisfaction and skill orientation	1/ 2	1/3	1	1/ 1.25	1/1.6
4	Use of IT and KM technologies	1/1.7	1/ 2.25	1.25	1	1/1.25
5	Use of QE and QM techniques	1/1.3	1/ 1.6	1.6	1.25	1

Sr. No.	Performance sub-criteria innovation and learning perspective	Weights
1	Accuracy of forecasting	0.240
2	Buyer supplier partnership level	0.318
3	Employee satisfaction and skill orientation	0.113
4	Use of IT and KM technologies	0.143
5	Use of QE and QM techniques	0.185

Max. Eigen Value (λ max) = 5.0024

Consistency Index = 0.0006 and Consistency Ratio = 0.0005.

Now, to find the impact of individual performance sub-criteria on the supply chain cyclic processes (alternatives in supply chain performance measurement framework).

Pair-wise comparison of alternatives with respect to performance criteria under financial perspective.

TABLE 5.96 Pair-Wise Comparison of Alternatives With Respect to Sub-Criteria (RROI) for Analysis of the Impact on Overall Supply Chain Performance.

Sr. No.	Supply chain cyclic processes (alternatives)	1	2	3	Weights
1	Procurement	1	1/ 2.2	1/1.4	0.216
2	Production	2.2	1	1.8	0.496
3	Distribution	1.4	1/1.8	1	0.288

A: Sub-criteria: Rate of return on investment.

Max. Eigen Value (λ max) = 3.00038

Consistency Index = 0.00019 and Consistency Ratio = 0.00032.

TABLE 5.97 Pair-Wise Comparison of Alternatives With Respect to Sub-Criteria (SCSI) for Analysis of the Impact on Overall Supply Chain Performance.

Sr. No.	Supply chain cyclic processes (alternatives)	1	2	3	Weights
1	Procurement	1	2.5	1	0.435
2	Production	1/ 2.5	1	1/1.25	0.219
3	Distribution	1	1.25	1	0.345

B: Sub-criteria: Supplier's cost-saving initiatives.

Max. Eigen Value (λ max) = 3.1

Consistency Index = 0.05 and Consistency Ratio= 0.086.

TABLE 5.98 Pair-Wise Comparison of Alternatives With Respect to Sub-Criteria (Inf. C. Cost) for Analysis of the Impact on Overall Supply Chain Performance.

Sr. No.	Supply chain cyclic processes (alternatives)	1	2	3	Weights
1	Procurement	1	1.25	1.6	0.409
2	Production	1/1.25	1	1.7	0.359
3	Distribution	1/1.6	1/1.7	1	0.233

C: Sub-criteria: Information carrying cost.

Max. Eigen Value (λ max) = 3.0016

Consistency Index = 0.0008 and Consistency Ratio = 0.0013.

TABLE 5.99 Pair-Wise Comparison of Alternatives With Respect to Sub-Criteria (Cost/h of Operation) for Analysis of the Impact on Overall Supply Chain Performance.

Sr. No.	Supply chain cyclic processes (alternatives)	1	2	3	Weights
1	Procurement	1	1/1.75	1/1.1	0.256
2	Production	1.75	1	2	0.483
3	Distribution	1.1	1/ 2	1	0.261

D: Sub-criteria: Cost per hour of operation.
Max. Eigen Value (λ max) = 3.0011
Consistency Index = 0.0055 and Consistency Ratio = 0.0094.

TABLE 5.100 Pair-Wise Comparison of Alternatives With Respect to Sub-Criteria (Inv. C. Cost) for Analysis of the Impact on Overall Supply Chain Performance.

Sr. No.	Supply chain cyclic processes (alternatives)	1	2	3	Weights
1	Procurement	1	1 /2	1.1	0.252
2	Production	2	1	2.5	0.529
3	Distribution	1/1.1	1/ 2.5	1	0.219

E: Sub-criteria: Inventory carrying cost.
Max. Eigen Value (λ max) =3.0036
Consistency Index = 0.0018 and Consistency Ratio = 0.0031.

TABLE 5.101 Pair-Wise Comparison of Alternatives With Respect to Sub-Criteria (Manufacturing Cost) for Analysis of the Impact on Overall Supply Chain Performance.

Sr. No.	Supply chain cyclic processes (alternatives)	1	2	3	Weights
1	Procurement	1	1/ 2	1.65	0.288
2	Production	2	1	2.5	0.523
3	Distribution	1/1.65	1/ 2.5	1	0.190

F: Sub-criteria: Manufacturing cost.
Max. Eigen Value (λ max) = 3.0178
Consistency Index = 0.0089 and Consistency Ratio = 0.015.
Pair-wise comparison of alternatives with respect to performance criteria under customer perspective.

TABLE 5.102 Pair-Wise Comparison of Alternatives With Respect to Sub-Criteria (QGM) for Analysis of the Impact on Overall Supply Chain Performance.

Sr. No.	Supply chain cyclic processes (alternatives)	1	2	3	Weights
1	Procurement	1	1/1.8	1/1.1	0.257
2	Production	1.8	1	2	0.487
3	Distribution	1.1	1/ 2	1	0.256

G: Sub-criteria: Quality of goods manufactured.
Max. Eigen Value (λ max) = 3.005
Consistency Index = 0.0025 and Consistency Ratio = 0.0043.

TABLE 5.103 Pair-Wise Comparison of Alternatives With Respect to Sub-Criteria (DP and LT) for Analysis of the Impact on Overall Supply Chain Performance.

Sr. No.	Supply chain cyclic processes (Alternatives)	1	2	3	Weights
1	Procurement	1	1/ 2.3	1/ 1.2	0.221
2	Production	2.3	1	2	0.517
3	Distribution	1.2	1/ 2	1	0.262

H: Sub-criteria: Delivery performance and lead time.
Max. Eigen Value (λ max) = 3.0004
Consistency Index = 0.0002 and Consistency Ratio = 0.00034.

TABLE 5.104 Pair-Wise Comparison of Alternatives With Respect to Sub-Criteria (Cust. Satisfaction) for Analysis of the Impact on Overall Supply Chain Performance.

Sr. No.	Supply chain cyclic processes (alternatives)	1	2	3	Weights
1	Procurement	1	1/ 2	1/ 1.5	0.219
2	Production	2	1	1.8	0.484
3	Distribution	1.5	1/1.8	1	0.297

I: Sub-criteria: Customer satisfaction.
Max. Eigen Value (λ max) = 3.018
Consistency Index = 0.009 and Consistency Ratio = 0.015.

TABLE 5.105 Pair-Wise Comparison of Alternatives With Respect to Sub-Criteria (EDPS) for Analysis of the Impact on Overall Supply Chain Performance.

Sr. No.	Supply chain cyclic processes (alternatives)	1	2	3	Weights
1	Procurement	1	1.2	1/1.5	0.303
2	Production	1/1.2	1	1/1.5	0.269
3	Distribution	1.5	1.5	1	0.429

J: Sub-criteria: Effectiveness of distribution planning schedules.
Max. Eigen Value (λ max) = 3.0074
Consistency Index = 0.0037 and Consistency Ratio = 0.0063.

TABLE 5.106 Pair-Wise Comparison of Alternatives With Respect to Sub-Criteria (CQT) for Analysis of the Impact on Overall Supply Chain Performance.

Sr. No.	Supply chain cyclic processes (alternatives)	1	2	3	Weights
1	Procurement	1	1.3	1/1.3	0.329
2	Production	1/1.3	1	1/1.4	0.269
3	Distribution	1.3	1.4	1	0.402

K: Sub-criteria: Customer query time.
Max. Eigen Value (λ max) = 3.0075
Consistency Index = 0.00375 and Consistency Ratio = 0.0064.

TABLE 5.107 Pair-wise comparison of alternatives With Respect to Sub-Criteria (RP&S) for Analysis of the Impact on Overall Supply Chain Performance.

Sr. No.	Supply chain cyclic processes (alternatives)	1	2	3	Weights
1	Procurement	1	1/ 2.25	1.8	0.287
2	Production	2.25	1	2	0.511
3	Distribution	1/1.8	1/ 2	1	0.202

L: Sub-criteria: Range of product and services.
Max. Eigen Value (λ max) = 3.04
Consistency Index = 0.02 and Consistency Ratio = 0.034.
Pair-wise comparison of alternatives with respect to performance criteria under internal business process perspective.

TABLE 5.108 Pair-Wise Comparison of Alternatives With Respect to Sub-Criteria (FMCN) for Analysis of the Impact on Overall Supply Chain Performance.

Sr. No.	Supply chain cyclic processes (alternatives)	1	2	3	Weights
1	Procurement	1	1/1.5	1.5	0.320
2	Production	1.5	1	2	0.460
3	Distribution	1/1.5	1/ 2	1	0.221

M: Sub-criteria: Flexibility to meet particular customer needs.
Max. Eigen Value (λ max) = 3.0032
Consistency Index = 0.0016 and Consistency Ratio = 0.0027.

TABLE 5.109 Pair-Wise Comparison of Alternatives With Respect to Sub-Criteria (TSCCT) for Analysis of the Impact on Overall Supply Chain Performance.

Sr. No.	Supply chain cyclic processes (alternatives)	1	2	3	Weights
1	Procurement	1	1/1.5	1.25	0.307
2	Production	1.5	1	1.5	0.429
3	Distribution	1/1.25	1/1.5	1	0.264

N: Sub-criteria: Total supply chain cycle time.
Max. Eigen Value (λ max) = 3.0098
Consistency Index = 0.0049 and Consistency Ratio = 0.0084.

TABLE 5.110 Pair-Wise Comparison of Alternatives With Respect to Sub-Criteria (PDCT) for Analysis of the Impact on Overall Supply Chain Performance.

Sr. No.	Supply chain cyclic processes (alternatives)	1	2	3	Weights
1	Procurement	1	1/ 2	1.25	0.263
2	Production	2	1	2.25	0.526
3	Distribution	1/1.25	1/ 2.25	1	0.211

O: Sub-criteria: Product development cycle time.
Max. Eigen Value (λ max) = 3.0004
Consistency Index = 0.0002 and Consistency Ratio = 0.00034.

TABLE 5.111 Pair-Wise Comparison of Alternatives With Respect to Sub-Criteria (Inf. C. Cost) for Analysis of the Impact on Overall Supply Chain Performance.

Sr. No.	Supply chain cyclic processes (alternatives)	1	2	3	Weights
1	Procurement	1	1/1.75	1/1.5	0.229
2	Production	1.75	1	2.1	0.487
3	Distribution	1.5	1/ 2.1	1	0.284

P: Sub-criteria: Capacity utilization.
Max. Eigen Value (λ max) = 3.08
Consistency Index = 0.04 and Consistency Ratio = 0.068.

TABLE 5.112 Pair-Wise Comparison of Alternatives With Respect to Sub-Criteria (PPCT) for Analysis of the Impact on Overall Supply Chain Performance.

Sr. No.	Supply chain cyclic processes (alternatives)	1	2	3	Weights
1	Procurement	1	1.25	1.05	0.361
2	Production	1/ 1.25	1	1.5	0.352
3	Distribution	1/ 1.05	1/1.5	1	0.287

Q: Sub-criteria: Planned process cycle time.
Max. Eigen Value (λ max) = 3.1
Consistency Index = 0.05 and Consistency Ratio = 0.086.

TABLE 5.113 Pair-Wise Comparison of Alternatives With Respect to Sub-Criteria (ECIQ) for Analysis of the Impact on Overall Supply Chain Performance.

Sr. No.	Supply chain cyclic processes (alternatives)	1	2	3	Weights
1	Procurement	1	1/ 2.5	1.5	0.260
2	Production	2.5	1	2	0.527
3	Distribution	1/1.5	1/2	1	0.214

R: Sub-criteria: Extent of cooperation to improve quality.
Max. Eigen Value (λ max) = 3.08
Consistency Index = 0.04 and Consistency Ratio = 0.068.
Pair-wise comparison of alternatives with respect to performance criteria under innovation and learning perspective.

TABLE 5.114 Pair-Wise Comparison of Alternatives With Respect to Sub-Criteria (AoF) for Analysis of the Impact on Overall Supply Chain Performance.

Sr. No.	Supply chain cyclic processes (alternatives)	1	2	3	Weights
1	Procurement	1	2	1.25	0.439
2	Production	1/ 2	1	1/1.25	0.238
3	Distribution	1/ 1.25	1.25	1	0.323

S: Sub-criteria: Accuracy of forecasting.
Max. Eigen Value (λ max) = 3.0128
Consistency Index = 0.0064 and Consistency Ratio= 0.011.

TABLE 5.115 Pair-Wise Comparison of Alternatives With Respect to Sub-Criteria (ECIQ) for Analysis of the Impact on Overall Supply Chain Performance.

Sr. No.	Supply chain cyclic processes (alternatives)	1	2	3	Weights
1	Procurement	1	2	1.5	0.460
2	Production	1/ 2	1	1/1.5	0.221
3	Distribution	1/ 1.5	1.5	1	0.319

T: Sub-criteria: Buyer supplier partnership level.
Max. Eigen Value (λ max) = 3.0028
Consistency Index = 0.00148 and Consistency Ratio= 0.0024.

TABLE 5.116 Pair-Wise Comparison of Alternatives With Respect to Sub-Criteria (ECIQ) for Analysis of the Impact on Overall Supply Chain Performance.

Sr. No.	Supply chain cyclic processes (alternatives)	1	2	3	Weights
1	Procurement	1	1/1.7	1.5	0.303
2	Production	1.7	1	2.05	0.479
3	Distribution	1/1.5	1/2.05	1	0.218

U: Sub-criteria: Use of IT and KM concepts.
Max. Eigen Value (λ max) = 3.0094
Consistency Index = 0.00474 and Consistency Ratio = 0.0081.

TABLE 5.117 Pair-Wise Comparison of Alternatives With Respect to Sub-Criteria (ECIQ) for Analysis of the Impact on Overall Supply Chain Performance.

Sr. No.	Supply chain cyclic processes (alternatives)	1	2	3	Weights
1	Procurement	1	1/ 2.5	1.25	0.248
2	Production	2.5	1	1.75	0.511
3	Distribution	1/1.25	1/1.75	1	0.241

V: Sub-criteria: Use of QE and QM techniques.
Max. Eigen Value (λ max) = 3.08
Consistency Index = 0.04 and Consistency Ratio = 0.068.

TABLE 5.118 Pair-Wise Comparison of Alternatives With Respect to Sub-Criteria (ECIQ) for Analysis of the Impact on Overall Supply Chain Performance.

Sr. No.	Supply chain cyclic processes (alternatives)	1	2	3	Weights
1	Procurement	1	1/ 2	1	0.256
2	Production	2	1	1.5	0.464
3	Distribution	1	1/1.5	1	0.280

W: Sub-criteria: Employee satisfaction and skill orientation.
Max. Eigen Value (λ max) = 3.018
Consistency Index = 0.009 and Consistency Ratio = 0.015.

TABLE 5.119 Importance of Supply Chain Cyclic Processes With Respect to Overall Supply Chain Performance Measurement.

Sr. No.	Supply chain cyclic processes (alternatives)	Final weight
1	Procurement	0.310
2	Production	0.422
3	Distribution	0.268

TABLE 5.120 Appendix E: Comparative Priorities of Supply Chain Cyclic Processes With Respect to Various Performance Measures (Sub-Criteria).

	Financial Perspective						Customer Perspective						Internal Business Process Perspective						Innovation & Learning Perspective				
Level of Sub-criteria	St	Tt	Cp	Op	Op	Op	Op	St	S, T, O	Tt	St	St	St	St	Tt	Op	Tt	S, T, O	St, Tt	St	S, T, O	S, T, O	S, T, O
Supply Chain Cyclic Processes	A	B	C	D	E	F	G	H	I	J	K	L	M	N	O	P	Q	R	S	T	U	V	W
Procurement	0.216	0.435	0.409	0.256	0.252	0.288	0.257	0.221	0.219	0.303	0.329	0.287	0.32	0.307	0.263	0.229	0.361	0.26	0.439	0.46	0.303	0.248	0.256
Production	0.496	0.219	0.356	0.483	0.529	0.523	0.487	0.517	0.484	0.269	0.269	0.511	0.46	0.429	0.526	0.487	0.352	0.527	0.238	0.221	0.479	0.511	0.464
Distribution	0.281	0.345	0.235	0.261	0.219	0.19	0.256	0.262	0.297	0.429	0.402	0.202	0.221	0.264	0.211	0.284	0.287	0.214	0.323	0.319	0.218	0.241	0.28
Max. Eigen Value (λ max)	3.0004	3.1	3.0016	3.0011	3.0036	3.0178	3.005	3.0004	3.018	3.0074	3.0075	3.04	3.0032	3.0098	3.0004	3.08	3.1	3.08	3.0128	3.0028	3.0094	3.08	3.018
Consistency Index	0.0002	0.05	0.0008	0.0055	0.0018	0.0089	0.0025	0.0002	0.009	0.0037	0.0038	0.02	0.0016	0.0049	0.0002	0.04	0.05	0.04	0.0064	0.00148	0.0047	0.04	0.009
Consistency Ratio	0.0003	0.086	0.0013	0.0094	0.0031	0.015	0.0043	0.00034	0.015	0.0063	0.0064	0.034	0.0027	0.0084	0.00034	0.068	0.086	0.066	0.011	0.0024	0.0081	0.068	0.015

KEYWORDS

- **supply chain performance measurement**
- **analytical hierarchy process**
- **data analysis**
- **performance criteria**
- **hierarchy and weights**
- **decision-making process**

CHAPTER 6

IMPLICATION AND CONTRIBUTION OF RESEARCH STUDY

ABSTRACT

This chapter highlights the implications and contributions of four case studies in the context of supply chain performance measurement (SCPM), providing insights into both the theoretical understanding and practical implementation of SCM and SCPM in the industry. It summarizes the general implications, contribution, theoretical value addition, practical value addition to the industry, and managerial implications. The implications of the four case studies shed light on various aspects of SCPM and its application in real-world scenarios. These implications encompass the identification of key performance indicators, the understanding of SC dynamics, the evaluation of SCPM frameworks, and the enhancement of SC efficiency and effectiveness. The contribution of the case studies lies in their ability to provide empirical evidence and practical insights that enhance the existing knowledge of SCM and SCPM. They offer a deeper understanding of the complexities and challenges faced by organizations in managing their SCs and measuring their performance. The theoretical value addition of the case studies lies in their contribution to the state of knowledge in SCM and SCPM. They offer new perspectives, frameworks, and methodologies that enrich the existing literature and provide researchers with a foundation for further exploration and advancement in the field. In terms of practical value addition, the case studies offer practical guidance and recommendations for industry professionals seeking to improve their SC performance measurement practices. They highlight

Supply Chain Performance Measurement in Textile Enterprises. Pranav C. Charkha, PhD, Santosh B. Jaju, PhD & Prasenjit Chatterjee, PhD (Authors.)

successful strategies, best practices, and lessons learned from real-world implementations, enabling organizations to enhance their SCPM capabilities and achieve better SC outcomes. Managerial implications arise from the findings and insights of the case studies. They provide guidance to managers and decision-makers in effectively implementing SCPM frameworks, aligning SC strategies with business goals, improving collaboration among SC partners, and driving continuous improvement in SC performance.

6.1 IMPLICATION AND CONTRIBUTION

6.1.1 IMPLICATION OF FOUR CASE STUDIES

6.1.1.1 FOR CASE COMPANY 1(5.5.1)

From Table 5.1, for the case of company 1, customer perspective is most important, followed by the internal business perspective. This shows that, company's supply chain performance measurement approaches justify their working philosophy, as the company has been serving the market since 1925, is conscious about customer expectations, and practises continual improvement in its internal business processes. From Table 5.2, supplier's cost saving initiative is worth considering as the most important, followed by information carrying costs and inventory carrying costs. This implies that for any company in any form of supply chain, either simple or an extended supply chain, initiatives to reduce costs at supplier's end are perceived to be very important, along with the information processing among the various shops of an industry or among the partners of supply chain. As the company produces a variety of textile products, especially suiting-shirting, hence large and varied type of raw material is required, which ultimately results in a large inventory, relating to cost associated with it. Table 5.3 reflects the ultimate aim of the company's supply chain to satisfy customer expectations by providing a variety of products and services with utmost standard of quality of good produced. It will be helpful in attaining the profit margin. The consciousness of industry towards supply chain is reflected in Table 5.4, where total supply chain cycle time is perceived to be most important, followed by the extent of cooperation in attaining quality benchmarks and achieving flexibility in order to meet customer needs. The company emphasizes its relation with immediate supplier or customer, followed by employing philosophies of continuous improvement using quality engineering/quality management techniques such as kaizen, TOC, ISO, etc., and efficient sharing of information and knowledge among all the levels of an industry or supply chain, as reflected in Table 5.5.

As shown in Table 5.29, among the relative comparisons of the three supply chain cycles, production of textile fabric or other material is perceived to be very important, followed by procurement of raw material and distribution to desired customer (wholesaler/retailer/direct to end user).

Table 5.30 (Appendix C) shows the comparative priorities of three supply chain cyclic processes with respect to various performance subcriteria. It shows that the financial perspective of the textile supply chain is mostly affected by the production technology or philosophy adopted by the textile industry, and also, we all know the textile production process is very lengthy and complex. Customer perspective is mostly a matter of concern for top management, which is reflected in the table as most of the strategic performance criteria have an impact for a customer aspect. It is clear that what services are required to be provided to customers is a decision of top management. Whereas internal business functions have a mixed blend impact of performance measures at the strategic, tactical, and operational levels. This is the place where the expectations of the customer are met by compiling the intentions of top management with those of managers, executives, and workers. Innovation and learning are functions to attain continual improvement, which must be initiated at every level of hierarchy and is reflected in the table as it has a chronological impact of strategic performance measures followed by tactical and operational measures.

This is the case of a company involved in manufacturing a number of products that belong to the suiting and shirting section, with a variety of quality benchmarks and thousands of retail outlets in the country.

6.1.1.2 FOR CASE COMPANY 2 (5.5.2)

From Table 5.31, for case company 2, again, customer perspective is most important, whereas, as compared to the earlier company here, it is followed by financial perspective. This shows that the company's supply chain performance measuring attributes believe in satisfying the customer by optimizing the finances incurred. This justifies their working philosophy, as the company is continuously producing the textile products thereby to minimize the cost involved in production. The company believes in the concept of make-to-stock. From Table 5.32, information carrying costs are perceived to be the most important, followed by suppliers' cost saving initiatives and inventory carrying costs. It shows that the company is well conscious of the information about the factors involved in the production, keeping the supply chain aspect in mind, like how much raw material is usable and what quantity is wasted. How many spindle set-ups are in operation, how much inventory level (raw material, fiber, yarn, and fabric), where to stock (warehouses), etc. This shows that the company relies on optimizing the cost along the supply chain by keeping the information of every cycle along the supply chain at a pinpoint location. As company produces single textile products, especially for bedding & home furnishing, hence large inventory is certain to be a matter of concern, but not that much critical as the company has constant demand for its product in southern India locally and in European countries globally. Table 5.33 shows that the company believes in good quality fabric (i.e., quality of goods manufactured), followed by the priority of delivering the products in the right condition and place at the right time (i.e., with good delivery performance and optimum lead time), hence, for this reason, the company's supply chain has incorporated warehouses at various locations in the country in order to satisfy customer expectations in a timely manner with right quality. The company believes in such a supply chain philosophy in order to sustain itself in the globalized market community of producing single textile products. The demand for the company's product is very high, which is a result of its policy of continual improvement, which is shown in Table 5.34 as the extent of cooperation to improve quality is perceived to be most important, followed by total supply chain cycle time and planned process cycle time by achieving efficient distribution planning schedules in order to meet customer needs. Company concern is more on recognizing the demand of the market by ensuring the forecasting process so as to produce the desired type of textile product, followed by enhanced use of technology facilitating information processing and knowledge management, thereby establishing better partnerships among the supply chain partners (immediate supplier and immediate customer), as emerged from Table 5.35.

As observed in Table 5.59, in the relative comparison of three supply chain cycles, procurement of the right quality raw material and other necessary additives is perceived to be very important, followed by complex production processes of converting raw material into finished fabric and distribution to warehouses/to the desired customer (wholesaler/retailer/direct to the end user).

Tables 5.36 to 5.58 elaborate on the pair wise comparison of supply chain cycles with respect to various performance sub-criteria. Whereas Table 5.60 (Appendix D) shows the comparative priorities of three supply chain cyclic processes with respect to various performance sub-criteria. In case company 2, the financial perspective of the textile supply chain is affected by the combination of procurement and production strategies. As a company's single textile product, the customer's perspective is mostly affected by how much to produce so as to reduce the cost of the product substantially by procuring the right quality of raw material. It is seen that this becomes a matter of concentration for top management, hence, impacted by strategic performance sub-criteria. Whereas internal business functions affecting procurement and production cycles of the supply chain with more importance on performance sub-criteria at the strategic, tactical levels. Innovation and learning are functions to attain continual improvement, and must be implemented at every stage of the supply chain, which must be started with the commitment of management. As observed from the table, the emphasis of performance sub-criteria under strategic is more followed by tactical, and then operational. One important thing that was observed is that as the company is producing a single textile product on a large scale, two major integrating supply chain cycles are procurement and production, which are perceived to be most important sequentially.

This is the case of a company involved in manufacturing unique textile products useful for bedding, home furnishing, etc., with quality standards and with various distribution warehouses located in the country.

6.1.1.3 FOR CASE COMPANY 3 AND 4

From Tables 5.61 and 5.91, for cases 3 and 4, the internal business process perspective has been very crucial, followed by the customer perspective in case of company 3 and the financial perspective in case of company 4, respectively. It shows that companies with a legacy of earlier experience in the textile sector are more concerned with fulfilling customer expectations from a customer perspective after its value addition business process, whereas companies with a smaller number of units are more focused with optimum utilization of finances after executing the internal business process to maximum extent. This reflects that internal business process perspective is of the utmost important for the textile supply chain for both companies. During pair wise comparison of performance measures under financial perspective (Tables 5.62 and 5.92), it is seen that inventory carrying cost perceived as most important followed by information carrying cost and supplier's cost saving initiatives, for customer perspective (Tables 5.63 and 5.93), quality of good manufactured is priority followed by customer satisfaction, and delivery performance and lead time, whereas in case of internal business process perspective (Tables 5.64 and 5.94), it is observed that extent of cooperation to improve quality received vital importance followed by planned process cycle time and total supply chain cycle time, and in case of innovation and learning perspective (Tables 5.65 and 5.95), buyer supplier partnership level perceived as important followed by accuracy of forecasting and use of quality engineering and quality management practices in industry to improve the performance of supply chain. From all these comparisons, it is evident that in both cases, companies are very identical in their supply chain performance measurement approaches. It is clear from both cases, companies 3 and 4 are believed to build cordial relationships with their immediate buyer/supplier and promote the use of best practices not only to improve productivity or production efficiency but also to be efficient in recognizing the pattern of customer demands, to increase the extent of cooperation for attaining quality products, optimizing the cycle time for production processes and for its supply chain, leading to the production of quality products/services, and to effectively minimizing delivery lead time to achieve the highest extent of customer satisfaction. Which ultimately follow expected financial returns and profit gains.

It is mentioned in the earlier chapter that both companies are serving the market with their synthetic yarn and unfinished fabric. Hence, their production processes are very complex. As a result of which efficient execution of all value addition activities during the manufacturing of yarn and fabric perceived critical concentration, followed by procurement of raw material so as to satisfy the production rate and quality requirement, and finally the distribution of these products to their extended counterparts, as observed in Tables 5.89 and 5.119 in the relative comparison of supply chain cycles.

Tables 5.66 to 5.88 and Tables 5.96 to 5.118 elaborate the pair wise comparison of supply chain cycles with respect to various performance sub-criteria for case companies 3 and 4, respectively. Whereas, Tables 5.90 (Appendix E) and 5.120 (Appendix F) show the comparative priorities of three supply chain cyclic processes with respect to various performance sub-criteria for case companies 3 and 4, respectively. In the case of companies 3 and 4, almost all the performance perspectives (criteria) for measuring the textile supply chain are dominantly affected by production strategies, followed by distribution and procurement cycles of equal importance. As both companies are producing only fiber and textile fabric for further application-based production, the production cycle is perceived to impact most of the performance sub-criteria, followed by distribution in case company 3, and procurement in case company 4.

6.2 GENERAL IMPLICATION

In this study, an important issue of supply chain management and supply chain performance measurement systems in textile industries in developing countries like India is focused on and discussed. The ultimate aim of the study is to give a set of generic performance measures as sub-criteria useful to evaluate the textile supply chain's performance. To attain this aim/ objective, hierarchical framework is finalized by executing a structured discussion with subject matter experts in the textile industry and then validated the framework for its applicability with four industries in our region. For this, the Delphi method is used to finalize the framework, which is hierarchical in nature, and the analytical hierarchy process (AHP) is used to analyze the ranking given by the personnel from the case companies.

The author initially conducted an extensive literature survey pertaining to supply chain management in the textile industry in India as well as the opportunities or possible areas for measuring the performance of the textile supply chain. Issues concerned with textile supply chain management resulting in efficient performance of its supply chain were reviewed. As the textile supply chain as being the process supply chain is significantly different from other discrete manufacturing companies and even from other process industries. Hence, the issues related to the textile supply chain need a special focus. The structured process of expert discussion (Delphi method) is explored, followed by the use and exploration of the multi-criteria decision making method. With the help of an extensive literature review from scholarly articles, academic books, practitioners' analyses, and the government's report, the current state of research and appropriate research gaps are identified.

Composite methodologies are utilized in the present study. Depending on the literature survey, the customized Delphi method is adopted to facilitate the author's best philosophical perception from the subject matter experts related to supply chain management, the textile industry from both industry and academia, and government association members and related statutory organization as well. The outcomes of the literature review and modified Delphi method were used to design. As the research in the textile industry in India is at a very budding stage, hence in order to develop an integrated supply chain performance measurement system, it is more suitable to identify the potential criteria, sub-criteria, and alternatives with the help we can gauge the performance of the textile supply chain. Selecting the best possible attributes/parameters out of multiple available, the best suitable method is to use multi-criteria decision making techniques, that is, analytical hierarchy process. The extensive literature review followed by a comprehensive face to face discussion with 15 subject matter experts enabled us to develop the supply chain performance measurement framework. These structured questionnaires are then put in front of the four industries (2 large scale industries and 2 small and 7 medium scale enterprises) to seek responses on the acceptability of the framework and its implementation in the textile industry. In all total, 78 employees (of various departments such as commercial, procurement, production, distribution, quality control, and human resource dept.) from the four identified industries submitted a completely filled questionnaire, whereas, 10 questionnaires were partially filled. On the other hand, 12 industry personnel were unable to fill the questionnaire.

At the end of the first and second stages of the Delphi method, a two-level hierarchical structure is suggested along with a proposal to include the third level in the hierarchy as an alternative in order to evaluate the contribution of alternatives to the overall supply chain performance. The proposal was discussed and finally included in the hierarchy. The need is to focus on holistic consideration of all the stakeholders in the industry to find out the impact on its supply chain performance. Hence, attributes such as financial, customer, internal business process, innovation, and learning are considered, which forms a comprehensive tool for performance measurement renowned as the balanced scorecard (BSC). These four BSC perspectives become the criteria for measuring supply chain performance, with 22 performance measures as sub-criteria to evaluate the performance at an even more micro level. The four perspectives viz., financial perspective, customer perspective, internal business process perspective, and innovation and learning perspective, are composed of 6, 6, 6, and 5 concrete sub-criteria at the second level, respectively. As per the proposal finalization, three supply chain cyclic processes that worked as coordinating processes between supply chain partners were added at the third level as an alternative. They are procurement, production, and distribution. Resultant framework for supply chain performance measurement in the textile industry is as shown in Figure 4.3.

During the discussion with subject matter experts in the Delphi technique, after finalizing the hierarchical structure, experts recommended that the further analysis should be done keeping in view industry-specific issues, and as we set the objective of our study, we were required to study the supply chain issues pertaining to specific industries from a performance measurement perspective. Hence, decided that instead of going for a survey for numerous industries, we identified four textile industries grouped into two categories; two large-scale industries and two belongs to small and medium-scale enterprises sector. All four industries are located in central India, in and around Nagpur.

The next stage is to evaluate the applicability of hierarchical structure in the four identified industries' supply chain contexts. The purpose of this step is to prioritize the selected criteria (financial perspective, customer perspective, internal business perspective, innovation, and learning perspective), sub-criteria (22(6, 6, 6, 5)) to find the crucial performance measures affecting the supply chain performance of the textile industry and the importance of alternative (supply chain cyclic processes—procurement, production, and distribution) responsible for contributing to supply chain performance as a whole. As the prioritization of performance measures seeks the concept of a multi-criteria decision-making approach. Hence, for ranking and finding priorities using pair wise comparison, the study used the AHP technique. The industry wise implication is already presented in an earlier chapter, whereas a comparative chart for various attributes at different levels of hierarchy is shown in Table 6.1.

After analyzing and prioritizing various performance measures in all four industries, it is concluded that supply chain performance measurement preferably depends on the following factors:

i. Size of the industry—Large or medium and small.
ii. Nature of industry—Composite or stand alone.
iii. Various types of raw material procured—Single (cotton) or multiple (synthetic fibers, etc.).
iv. Complexity of production process (Only ginning only, spinning, or only weaving or complete up to processing/furnishing).
v. Production philosophy (with customization/without).
vi. Efforts of continual improvement initiatives for quality and productivity enhancement.
vii. Rate and quality of information sharing and processing.
viii. Effective storage and retrieval and distribution warehouses.
ix. Involvement of all the stakeholders.
x. Integration of supply chain partners.

6.3 CONTRIBUTION

6.3.1 THEORETICAL VALUE ADDITION-STATE OF KNOWLEDGE OF SCM AND SCPM

Keeping in view the theoretical aspect, this is the first of its kind of work conducted in the Indian textile industry. The study identifies the issues in the textile supply chain and aligns the objectives of the study with these issues. It facilitates the comprehensive nature of practices of supply chain performance measurement in textile industries by including the framing of the hierarchical structure of supply chain performance and latter applicability validation in four different types of industries broadly classified as large scale and small and medium enterprises. The entire philosophy of textile industry production was discussed from the viewpoint of supply chain performance measurement. Work was conducted earlier also, but the attempt in this study was unique by combining the Delphi method for subject matter experts used for framework designing with AHP analysis of identified industries for prioritization of performance measures and finding the impact contribution of supply chain cyclic processes. As a result, it

TABLE 6.1 Summary of Criteria, Sub-criteria, and Alternatives for their Priority Importance (as per Weights Achieved).

		Priority weights criteria, sub-criteria for case companies			
Sr. No	**BSC performance perspectives**	**Case Company-1**	**Case Company-2**	**Case Company-3**	**Case Company-4**
1	BSC performance perspective	Customer perspective (0.394)	Customer perspective (0.363)	Internal business process perspective (0.366)	Internal business process perspective (0.401)
		Internal business process perspective (0.285)	Financial perspective (0.257)	Customerperspective (0.262)	Financial perspective (0.248)
		Innovation and learning perspective (0.205)	Internal business process perspective (0.216)	Financial perspective (0.201)	Customer perspective (0.192)
		Financial perspective (0.116)	Innovationanand learning perspective (0.163)	Innovation and learning perspective (0.170)	Innovation and learning perspective (0.160)
2	Financial perspective	Supplier's cost saving initiative (0.278)	Information carrying cost (0.285)	Inventory carrying cost (0.273)	Inventory carrying cost (0.270)
		Information carrying cost (0.235)	Supplier's cost saving initiative (0.224)	Information carrying cost (0.219)	Information carrying cost (0.228)
		Inventory carrying cost (0.182)	Inventory carrying cost (0.165)	Supplier's cost saving initiative (0.219)	Supplier's cost saving initiative (0.171)
3	Customer perspective	Customer satisfaction (0.320)	Quality of goods manufactured (0.280)	Quality of goods manufactured (0.283)	Quality of goods manufactured (0.277)
		Range of product and services (0.217)	Delivery performance and lead time (0.207)	Customer satisfaction (0.202)	Customer satisfaction (0.218)
		Quality of goods (0.154) manufactured	Customer satisfaction (0.181)	Delivery performance and lead time (0.166)	Delivery performance and lead time (0.165)

TABLE 6.1 *(Continued)*

		Priority weights criteria, sub-criteria for case companies			
Sr. No	**BSC performance perspectives**	**Case Company-1**	**Case Company-2**	**Case Company-3**	**Case Company-4**
4	Internal business Process perspective	Total supply chain cycle time (0.290)	Extent of cooperation to improve quality (0.276)	Extent of cooperation to improve quality (0.292)	Extent of cooperation to improve quality (0.275)
		Extent of cooperation to improve quality (0.230)	Total supply chain cycle time (0.201)	Planned process cycle time (0.205)	Planned process cycle time (0.206)
		Flexibility to meet particular customer's needs (0.169)	Planned process cycle time (0.168)	Total supply chain cycle time (0.172)	Total supply chain cycle time (0.170)
5	Innovation and learning perspective	Buyer supplier partnership level (0.326)	Accuracy of forecasting (0.302)	Buyer-supplier partnership level (0.316)	Buyer-supplier partnership level (0.318)
		Use of QE and QM techniques (0.245)	Use of IT and KM technologies (0.236)	Accuracy of forecasting (0.237)	Accuracy of forecasting (0.240)
		Use of IT and KM technologies (0.185)	Buyer-supplier partnership level (0.190)	Use of QE and QM techniques (0.180)	Use of QE and QM techniques (0.185)
6	Contribution of supply chain cyclic processes towards overall supply chain performance of textile industry	Production (0.435)	Procurement (0.377)	Production (0.409)	Production (0.464)
		Procurement (0.302)	Production (0.331)	Procurement (0.301)	Distribution (0.280)
		Distribution (0.263)	Distribution (0.292)	Distribution (0.290)	Procurement (0.256)

filled a gap in the existing state of the art of research in supply chain performance measurement for the textile industry, specially in developing countries like India.

Yet another uniqueness of the study is the construction of a supply chain performance measurement system in the textile industry, which is perspective-based taking care of stakeholders and finding the impact of supply chain cyclic processes between integrating supply chain partners in a more holistic and balanced manner. The ultimate objective of SCM is to have a system approach that considers the chain partners as a single entity, instead of fragmented units, which is really a big issue in textile supply chain performance measurement systems. It was very rare, that the research in the supply chain performance measurement was holistic and with balanced of all in one.

6.3.2 PRACTICAL VALUE ADDITION TO INDUSTRY

The biggest contribution of this study is the application of the BSC perspective, thereby considering stakeholder aspects. Yet another important practical contribution of this study is the comprehensive set of performance measures considering the quantitative and qualitative nature of sub-criteria for evaluating textile supply chain performance classified under each BSC perspective. Also, the performance measures considered are selected to cover the three decision making levels (strategic, tactical, and operational).

The research is different from the existing studies in the sense that it has considered the perceptions and beliefs of industry personnel working in six different departments over the philosophy of operation management, etc., ensuring the effectiveness of supply chain of the two broad categories of the textile industry (two large scale and two small and medium enterprises) as compared to other studies where pre-assumed notions were tested for their significance and a few industry experts were discussed for the solution.

Another important focus of the study is the priority weight of first- and second-level criteria and sub-criteria, respectively, in textile supply chain. In this research, instead of conducting a survey questionnaire of industries, expert personnel from key functional departments of the textile industry were surveyed as attributes responsible for implementing strategies pertaining to supply chain management as well as monitoring it in an effective and efficient manner. The respondents include industry experts from companies in the textile supply chain, for example, garment manufacturers, fabric manufacturers, yarn manufacturers, distributors, and raw material suppliers. The result is synthesized from their genuine responses/opinions. Hence, can definitely having a practical base.

Yet another important contribution that differs from previous studies is the consideration of supply chain cyclic processes (procurement, production, and distribution) as alternatives to find out, which of the processes affects the supply chain's performance more. The selection of these alternatives is based on expert opinion of the Delphi method. These are the integration processes between different supply chain partners. Supply chain cyclic processes were pair wise compared to find out the impact of all BSC perspectives. Their priorities were calculated for every submeasure in order to estimate the final weight, revealing their severity towards supply chain performance.

6.4 MANAGERIAL IMPLICATION

The study has achieved the priority ranking of various sub-criteria under different BSC perspectives, which are tested and verified for measuring textile supply chain performance. Other than typical quantitative criteria, qualitative criteria such as employing IT/KM technologies and using QE/QM techniques are also included. It helps us to give a more comprehensive and qualitative nature for evaluation.

A hierarchical framework has been formulated using the BSC perspective for evaluating textile SC. An AHP methodology is used to establish pair-wise comparisons, determine the relative weights of BSC perspectives, and determine the relative importance of sub-criteria under each perspective. To gauge the performance of textile SC with reference to the strategy followed for improving efficiency and effectiveness, three strategic process divisions have been considered to find out the relative importance of criteria and sub-criteria in different divisions. This enables companies to explore their focus to improve SC effectiveness. This study has particular relevance to textile SC in India, where the government is taking initiative in increasing the textile productivity in the country by establishing integrated textile parks (i.e., ginning, spinning, weaving, and processing at one place), and hence it becomes important for enterprises to have controlled as well as more effective and efficient SCM.

Current study is based on respondents from four textile industries in India. Perceptions and views in other countries could vary in results. The study presented a model that is generic in nature, hence wouldn't be changed. But the criteria were selected based on the practices and ground realities in the Indian context, which may vary to suit global competencies in the textile SCM aspect.

KEYWORDS

- **supply chain performance measurement**
- **case studies**
- **theoretical value addition**
- **practical value addition**
- **managerial implications**

REFERENCES

Abdul-Hamid, Y. T. The Analytical Hierarchy Process Approach to the Choice of Manufacturing Plant Layout. *Proc. Inst. Mech. Eng. B J. Eng. Manuf.* **1999,** *213* (B4), 397–406.

Adams, S. M., Sarkis, J., Liles, D. The Development of Strategic Performance Measures. *Eng. Manag. J.* **1995,** *7* (10), 24–32.

Agrawal, N.; Smith, S. A.; Tsay, A. A. Multi-Vendor Sourcing in a Retail Supply Chain. *Int. J. Prod. Oper. Manag.* **2002,** *11* (2), 157–182.

Angerhofer, B. J.; Angelides, M. C. A Model and a Performance Measurement System for Collaborative Supply Chains. *Decis. Support Syst.* **2006,** *42* (1), 283–301.

Aramyan, L. H.; Lansink, A. O.; Vorst, J.; Kooten, O. Performance Measurement in Agri-food Supply Chains: A Case Study. *Supply Chain Manag. Int. J.* **2007,** *12* (4), 304–315.

Beamon, B. M. Supply Chain Design and Analysis: Models and Methods. *Int. J. Prod. Econ.* **1998,** *55* (3), 281–294. Beamon, B. M., Measuring Supply Chain Performance. *Int. J. Oper. Prod. Manag.* **1999,** *19* (3), 275–292.

Beamon, B. M.; Chen, V. C. P. Performance Analysis of Conjoined Supply Chains. *Int. J. Prod. Res.* **2001,** *39* (914), 3195–3218.

Bergquist, K.; Abesysekera, J. QFD – A Means for Developing Usable Products. *Int. J. Ind. Ergon.* **1996,** *18*, 269–275.

Berrah, L.; Cliville, V. Towards an Aggregation Performance Measurement System Model in a Supply Chain Context. *Comput. Ind.* **2007,** *58*, 709–719.

Berry, D.; Towill, D. R.; Wadsley, N. Supply Chain Management in the Electronics Products Industry. *Int. J. Phys. Distrib. Logist. Manag.* **1994,** *24* (10), 20–32.

Bhagwat, R.; Sharma, M. K. Performance Measurement of Supply Chain Management: A Balanced Scorecard Approach. *Comput. Ind. Eng.* **2007a,** *53*, 43–62.

Bhagwat, R.; Sharma, M. K. Performance Measurement of Supply Chain Management Using the Analytical Hierarchy Process. *Prod. Plan. Control* **2007b,** *18* (8), 666–680.

Bheda, R.; Narag, A. S.; Singla, M. L. Apparel Manufacturing: A Strategy to Productivity Improvement. *J. Fash. Mark. Manag.* **2003,** *7* (1), 12–22.

Bigliardi, B.; Bottani, E. Performance Measurement in the Food Supply Chain: A Balanced Scorecard Approach. *Facilities* **2010,** *28* (5/6), 249–260.

Bourne, M.; Neely, A.; Mills, J.; Platts, K. Implementing Performance Measurement System: A Literature Review. *Int. J. Bus. Perform. Manag.* **2003,** *5* (1), 1–24.

Bremser, W. G.; Barsky, N. P. Utilizing the Balanced Scorecard for R&D Performance Measurement. *R&D Manag.* **2004,** *34* (3), 229–238.

Browne, J.; Devlin, J.; Rolstadas, A.; Andersen, B. Performance Measurement: The ENAPS Approach. *Int. J. Bus. Transform* **1998,** *1* (2), 73–84.

Bruce, M.; Daly, L.; Towers, N. Lean or Agile: A Solution for Supply Chain Management for Textile and Clothing Industry? *Int. J. Oper. Prod. Manag.* **2004,** *24* (2), 151–170.

Busi, M.; Bititci, S. Collaborative Performance Measurement: - Present Gaps & Future Research. *Int. J. Prod. Perform. Manag.* **2006,** *55* (1), 7–25. DOI: 10.1108/17410400610635471

Calantone, R. J. Using Analytical Hierarchy Process in New Product Screening. *Int. J. Prod. Innov. Manag.* **1999,** *16*, 65–76.

Cai, J.; Liu, X.; Xiao, Z.; Liu, J. Improving Supply Chain Performance Management: A Systematic Approach to Analyzing Iterative KPI Accomplishment. *Decis. Support Syst.* **2009,** *46*, 512–521.

Chae, B. Developing Key Performance Indicators for Supply Chain: An Industry Perspective. *Supply Chain Manag. Int. J.* **2009,** *14* (6), 422–428.

Chan, F. T. S. Performance Measurement in a Supply Chain. *Int. J. Adv. Manuf. Technol.* **2003,** *21*, 534–548.

Chan, F. T. S.; Qi, H. J. An Innovative Performance Measurement Method for Supply Chain Management. *Supply Chain Manag. Int. J.* **2003,** *8* (3-4), 209–223.

Chan, F. T. S.; Qi, H. J.; Chan, H. K.; Lau, H. C. W.; Ip, R. W. L. A Conceptual Model of Performance Measurement for Supply Chains. *Manag. Decis.* **2003,** *41* (7), 635–642.

Chandra, C.; Kumar, S. An Application of System Analysis Methodology to Manage Logistic in a Textile Supply Chain. *Supply Chain Manag.* **2000,** *5* (5), 234–244.

Charkha, P. G.; Jaju, S. B. Supply Chain Performance Measurement System: An Overview. *Int. J. Bus. Perf. Supply Chain Modeling* **2014 a,** *6* (1), 40–60.

Charkha, P. G.; Jaju, S. B. Designing Innovative Framework for Supply Chain Performance Measurement in Textile Industry. *Int. J. Logist. Syst. Manag.* **2014b,** *6* (1), 216–230.

Chia, A.; Goh, M.; Hum, S. H. Performance measurement in supply chain entities: balanced scorecard perspective. *Benchmarking Int. J.* **2009,** *16* (5), 605–620.

Chopra, S.; Meindl, P. In *Supply Chain Management: Strategy, Planning, and Operation*, 3rd ed.; Pearson Education: Upper Saddle River, NJ, 2007.

Christopher, M.; Juttner, U. Developing Strategic Partnerships in the Supply Chain: A Practitioner Perspective. *Eur. J. Purch. Supply Chain Manag.* **2000,** *2* (6), 117–127.

CRISIL Report- Customised Research Bulletin for Textile Industry, January 2012.

CRISIL Report (Infrastructure Advisory)- National Manufacturing Competitiveness Council, Enhancing Competitiveness of Indian Manufacturing Industry: Assistance in Policy Making, March 2009.

Czinkota, M. R.; Ronkainen, I. A. In *International Marketing*; Thomson South Western, 2007.

Dangayach, G. S.; Deshmukh, S. G. Evidence of Manufacturing Strategies in Indian Industry: A Survey. *Int. J. Prod. Econ.* **2003,** *83* (3), 279–298.

Dey, P. K. Benchmarking Project Management Practices of Caribbean Organizations Using Analytic Hierarchy Process. *Benchmarking Int. J.* **2002,** *9* (4), 326–356.

Dey, P. K. Integrated Approach to Project Selection Using Multiple Attribute Decision-Making Technique. *Int. J. Prod. Econ.* **2006,** *103* (1), 90–103.

Dey, P. K.; Ramcharan, E. K. Analytic Hierarchy Process Helps Select Site for Limestone Quarry Expansion in Barbados. *J. Environ. Manag.* **2008,** *88* (4), 1384–1395.

Dey, P. K.; Tabucanon, M. T.; Ogunlana, S. O. Planning for Project Control Through Risk Analysis: A Case of Petroleum Pipeline Laying Project. *Int. J. Proj. Manag.* **1994,** *12* (1), 23–33.

Dixon, J. R.; Nanni, A. J.; Vollman, T. E. In *The New Performance Challenge: Measuring Operations for World Calls Competition*; Dow-Jones Irwin: Homewood, IL, 1990.

Dweiri, F.; Al-oqla, F. M. Material Selection Using Analytical Hierarchy Process. *Int. J. Comput. Appl. Technol.* **2006,** *26* (4), 182–189.

Dweiri, F.; Meirer, F. A. Application of Fuzzy Decision-Making in Facility Lay Out Planning. *Int. J. Prod. Res.* **1996,** *34* (11), 3201–3225.

Farley, G. A. Discovering Supply Chain Management: a Roundtable Discussion. *APICS – The Performance Advantage* **1997,** *7* (1), 38–39.

Flynn, B. B.; Huo, B.; Zhao, X. The Impact of Supply Chain Integration on Performance: A Contingency and Configuration Approach. *J. Oper. Manag.* **2010,** *28* (1), 58–71.

Folan, P.; Browne, J. A Review of Performance Measurement: Towards Performance Management. *Comput. Ind.* **2005,** *56*, 663–680.

Forza, C.; Vinelli, A. Quick Response in Textile-Apparel Industry and the Support of Information Technologies. *Integr. Manuf. Syst.* **1997,** *8* (3), 125–136

Gammelgaard, B.; Vesth, H. In *The SCOR Model – A Critical Review*, Proceedings of Operations Management as a Change Agent Conferences. INSEAD, 2004; pp 233–241.

Goh, M.; Pinaikul, P. Logistics Management Practices and Development in Thailand. *Logist. Inf. Manag.* **1998,** *11* (6), 359–369.

Gunasekaran, A.; Kobu, B. Performance Measures and Metrics in Logistics and Supply Chain Management: A Review of Recent Literature (1995–2004) for Research and Applications. *Int. J. Prod. Res.* **2007,** *45* (12), 2819–2840.

Gunasekaran, A.; Ngai, N. T. Eric Build to Order Supply Chain Management: A Literature Review and Framework for Development. *Int. J. Oper. Manag.* **2007,** *23* (5), 423–451.

Gunasekaran, A.; Patel, C.; McGaughey, R. E. A Framework for Supply Chain Performance Measurement. *Int. J. Prod. Econ.* **2004,** *87* (3), 333–347.

Gunasekaran, A.; Patel, C.; Tirtiroglu, E. Performance Measures and Metrics in a Supply Chain Environment. *Int. J. Oper. Prod. Manag.* **2001,** *21* (1-2), 71–87.

Hair, J. F. Jr. In *Multivariate Data Analysis: With Reading* (Book), Prentice-Hall Inc.: USA, 1995. ISBN: -0-02-349020-9

Handfield R. B.; Nichols E. L. In *Introduction to Supply Chain Management*; Prentice Hall: New Jersey, 1999.

Harland, C. M.; Lamming, R. C.; Cousins, P. D. Developing the Concept of Supply Strategy. *Int. J. Oper. Prod. Manag.* **1999,** *19* (7), 650–673.

Harwick, T. Optimal Decision-Making for the Supply Chain. *APICS – The Performance Advantage* **1997,** *7* (1), 42–44.

Helmer O. In *Looking Forward: A Guide to Future Research*; Sage Publication: Beverly Hills, 1983.

Hervani, A. A.; Helms, M. M.; Sarkis, J. Performance Measurement for Green Supply Chain Management. *Benchmarking Int. J.* **2005,** *12* (4), 330–353.

Holmberg, S. A System Perspective in Supply Chain Measurement. *Int. J. Phys. Distrib. Log. Manag.* **2000,** *30* (10), 847–868.

Huan, S. H.; Sheoran, S. K.; Wang, G. A Review and Analysis of Supply Chain Operations Reference (SCOR) Model. *Supply Chain Manag. Int. J.* **2004,** *9* (1), 23–29.

Huang, S. H.; Sheoran, S. K.; Keskar, H. Computer-Assisted Supply Chain Configuration Based on Supply Chain Operations Reference (SCOR) Model. *Comput. Ind. Eng.* **2005,** *48*, 377–394.

Hudson, M.; Lean, J.; Smart, P. A. Improving Control Through Effective Performance Measurement in SMEs. *Prod. Plan. Control,* **2001,** *12* (8), 804–813.

Hwang, Y. D.; Lin, Y. C.; Lyu Jr., J. The Performance Evaluation of SCOR Sourcing Process – The Case Study of Taiwan's TFT-LCD Industry. *Int. J. Prod. Econ.* **2008,** *115*, 411–423.

Jagdev, H.; Brennan, A.; Browne, J. In *Strategic Decision Making in Modern Manufacturing*; Kluwer Academic Publishers: Massachusetts, MA, 2004.

Jammernegg, W.; Reiner, G. Performance Improvement of Supply Chain Processes by Coordinated Inventory and Capacity Management. *Int. J. Prod. Econ.* **2007,** *108*, 83–190.

Jones, R. M. In *The Apparel Industry*; Blackwell Science: Oxford, 2002.

Kaplan, R. S.; Norton, D. P. The Balanced Scorecard: Measures that Drive Performance. *Harvard Bus. Rev.* **1992,** *70* (1), 71–99.

Kaplan, R. S.; Norton, D. P. Using the Balanced Scorecard as a Strategic Management System. *Harvard Bus. Rev.* **1996,** *74* (1), 75–85.

Kennerley, M.; Neely, A. A Framework of the Factors Affecting the Evolution of Performance Measurement Systems. *Int. J. Oper. Prod. Manag.* **2002,** *22* (11), 1222–1245.

Khan, M. U. A Comparative Study on Distribution Channels of Textile & Garment Industry in Pakistan & Vietnam: Role of Intermediaries, Faculty of Economic Sciences, Communication & IT, 2010.

Kosela, L. In *Management of Production Construction: A Theoretical View*, Proceedings of the 7th Annual Conference of the International Group for Lean Construction IGLC-7; Berkeley, 1999; 241–252.

Kurien, G. P.; Qureshi, M. N. Performance Measurement System for Green Supply Chain Using Modified Balanced Scorecard and Analytical Hierarchy Process. *Sci. Res. Essays* **2012,** *7* (36), 3149–3161.

Lai, K. H.; Ngai, E. W. T.; Cheng, T. C. E. Measures for Evaluating Supply Chain Performance in Transport Logistics. *Transp. Res. Part E Log. Transp. Rev.* **2002,** *33* (6), 439–456.

Lambert, D. M.; Garcia-Dastugue, S. J.; Croxton, K. L. An Evaluation of Process-Oriented Supply Chain Management Frameworks. *J. Bus. Log.* **2005,** *26* (1), 25–51.

Laosirihongthong, T.; Dangayach, G. S. New Manufacturing Technology Implementation: A Study of the Thai Automotive Industry. *Prod. Plan. Control* **2005,** *16* (3), 263–272.

Lapide, L. What About Measuring Supply Chain Performance? *ASCET* **2000,** *2* (15), 287–297.

Lee, H. L.; Billington, C. Managing Supply Chain Inventory: Pitfalls and Opportunities. *Sloan Manag. Rev.* **1992,** *33* (3), 65–73.

Lee, Y.; Kincade, D. H. US Apparel Manufacturers-Company Characteristic Difference Based on SCM Activities. *J. Fash. Mark. Manag.* **2003,** *7* (1), 31–48.

Leung, S. Y. S. World Class Apparel-Sourcing Enterprises and the Restructuring of Existing Global Supply Chains. *J. Text. Inst. Part 2* **2000,** *91* (2), 73–93.

Lin, Y.; Wang, Y.; Yu, C. Investigating the Drivers of the Innovation in Channel Integration and Supply Chain Performance: A Strategy Orientated Perspective. *Int. J. Prod. Econ.* **2010,** *127* (2), 320–332.

Linstone, H.; Turoff, M. In *Introduction to Delphi Method: Techniques and Application*; Linstone & Turoff, Eds.; Wesley Publication Company: London: Addison, 1975.

Lohman, C.; Fortuin, L.; Wouters, M. Designing a Performance Measurement System: A Case Study. *Eur. J. Oper. Res.* **2004,** *156* (2), 267–286.

Lummus, R. R.; Vokurka, R. J. Defining Supply Chain Management: A Historical Perspective and Practical Guidelines. *Ind. Manag. Data Syst.* **1999,** *99* (1), 11–17.

McMullen, A. Supply Çhain Management Practices in Asia Pacific Today. *Int. J. Phys. Distrib. Log.* **1996,** *26* (10), 79–95.

Mentzer, J.; DeWitt, W.; Keebler, J.; Min, S.; Nix, N.; Smith, C.; Zacharia, Z. Defining Supply Chain Management. *J. Bus. Log.* **2001,** *22* (2), 1–25.

Mohanty, R. P.; Deshmukh, S. G. In *Supply Chain Management: Theories & Practices*; Dreamtech Publisher: WileyIndia, 2011. ISBN: 9788177221916.

Morgan, J.; Monczka, R. M. Alliances for New Products. *Purchasing* **1995,** *118* (1), 103–109.

Morgan, J.; Monczka, R. M. Supplier Integration: A New Level of Supply Chain Management. *Purchasing* **1996,** *120* (1), 110–113.

Neely, A.; Gregory, M.; Platts, K. The Performance Measurement Revolution: Why Now and What Next? *Int. J. Oper. Prod. Manag.* **1999,** *19* (2), 205–228.

New, S.; Ramsay, J. A Critical Appraisal of Aspects of the Lean Approach. *Eur. J. Purch. Supply Manag.* **1997,** *3* (2), 93–102.

Nowell, H. Market Competitiveness in Global Supply Chain: Examination of Supply Chain Configuration-Master's Thesis, North Carolina University, 2005.

Oliver, R. K.; Webber, M. D. Supply Chain Management: Logistics Catches up with the Strategy. In *Logistics, The Strategic Issue*; Christopher, M. G., Ed.; Chapman & Hall: London, 1992.

Otto, A.; Kotza, H. Does Supply Chain Management Really Pay? Six Perspectives to Measure the Performance of Managing a Supply Chain. *Eur. J. Oper. Res.* **2003,** *144*, 306–320.

Pasutham, Asawin. "Supply chain performance measurement framework: Case studies on the Thai manufacturers." PhD diss., Aston University, 2012.

Perry, M.; Sohal, A. Improving Service Quality Within the Supply Chain: An Australian Study. *Total Qual. Manag.* **1999,** *10* (4/5), 673-678.

Powlowski, S. D. The Delphi Method as a Research Tool: An Example, Design, Consider & Application. *J. Inf. Manag.* **2004,** *42* (1), 15–29. DOI: 10.1016/j.im.2003.11.002

Pratabaraj, J.; Wesley, R. J.; Melinda, A. C. Analysis on Supply Chain Performance in the Textile Industry with Tirupur as the Sampling Frame. *J. Commer. Manag. Res.* **2011**, *1* (3), 15–24. — ISSN 2249-0422,

Robb, D. J.; Xie, B.; Arthanari, T. Supply Chain and Operations Practice and Performance in Chinese Furniture Manufacturing. *Int. J. Prod. Econ.* **2008,** *112*, 683–699.

Rodriguez, R. R.; Saiz, J. J. A.; Bas, A. O. Quantitative Relationships Between Key Performance Indicators for Supporting Decision-Making Processes. *Comput. Ind.* **2009,** *60*, 104–113.

Ross, D. F. Competing Through Supply Chain Management; Chapman & Hall: New York, 1998.

Rouse, P.; Putterill, M. An Integral Framework for Performance Measurement. *Manag. Decis.* **2003,** *41* (8), 791–805.

Saad, M.; Jones, M.; James, P. A Review of the Progress Towards the Adoption of Supply Chain Management (SCM) Relationships in Construction. *Eur. J. Purch. Supply Manag.* **2002,** *8*, 173–183.

Saaty, T. L. An Exposition of the AHP in Reply to the Paper 'Remarks on the Analytic Hierarchy Process'. *Manag. Sci.* **1990,** *36* (3), 259–268.

Sauls, J. Excellence in Sourcing, Apparel Magazine Supply **2007,** *48* (12), 2–9.

Shepherd, C.; Gunter, H. Measuring Supply Chain Performance: Current Research and Future Directions. *Int. J. Product. Perform. Manag.* **2006,** *55* (3/4), 242–258.

Simatupang, T. M.; Sridharan, R. Design for Supply Chain Collaboration. *Bus. Process Manag. J.* **2008,** *14* (3), 401–418.

Slack, N.; Chambers, S.; Johnston, R. In *Operations Management*, 3rd ed.; Prentice Hall: Eaglewood Cliff, NJ, 2001.

Sople, V. V. In *Supply Chain Management: Text & Cases*; Pearson Education Publisher, 2012. ISBN: - *8131760995*

Spekman, R. E.; Kamauff, J. W. Jr.; Myhr, N. An Empirical Investigation into Supply Chain Management: A Perspective on Partnerships. *Int. J. Phys. Distrib. Log. Manag.* **1998,** *28* (8), 630–650.

Stein, T.; Sweat, T. In *Killer Supply Chains*; Information Week, 1998; vol 708 (9), pp 36–46.

Steven, G. C. Integrating the Supply Chain. *Int. J. Phys. Distrib. Log. Manag.* *19* (8), 3–8.

Swaninathan, J. M. In *Managing Supply Chain Operations in India-Pitfalls & Opportunities*. Building Supply Chain Excellence in Emerging Economies; Springer Science + Business Media, LLC, 2006.

Tan, K. C.; Handfield, R. B.; Krause, D. R. Enhancing Firm's Performance Through Quality and Supply Base Management: An Empirical Study. *Int. J. Prod. Res.* **1998a,** *36* (10), 13–37.

Tan, K. C.; Kannan, V. R.; Handfield, R. B. Supply Chain Management: Supplier Performance and Firm Performance. *Int. J. Purch. Mater. Manag.* **1998b,** *34* (3), 2–9.

Tan, K. C.; Lyman, S. B.; Wisner, J. D. Supply Chain Management: A Strategic Perspective. *Int. J. Operat. Prod. Manag.* **2002,** *22* (6), 614–631.

Thakkar, J.; Kanda, A.; Deshmukh, S. G. Supply Chain Performance Measurement Framework for Small and Medium Scale Enterprises. *Benchmarking Int. J.* **2009,** *16* (5), 702–723.

The Supply Chain Council. About Supply Chain Council [Online] 2009a. http://supply-chain.org/about

The Supply Chain Council. What is SCOR? [Online] 2009b. http://supply-chain.org/scor

Theeranuphattana, A.; Tang, J. C. S. A Conceptual Model of Performance Measurement for Supply Chains Alternative Considerations. *J. Manuf. Technol. Manag.* **2008,** *19* (1), 125–148.

Turner, T. J.; Bititci, U. S.; Nudurupati, S. S. Implementation and Impact of Performance Measures in Two SMEs in Central Scotland. *Prod. Plan. Control* **2005,** *16* (2), 135–151.

Ukalkar, S. Strategic Procurement Management for Competitive Advantage; Oxford University Press: London, 2000.

Von Hipple, E. Co-operation Between Rivals: Informal Know how Trading. *Res. Policy* **1986,** *16* (5), 21–28.

Vrijhoef, R.; Koskela, L. The Four Roles of Supply Chain Management in Construction. *Eur. J. Purch. Supply Manag. Supply Chain Manag. Const. Sp. Issue* **2000,** *6*, 169–178.

Waller, D. In *Operations Management: A Supply Chain Approach,* 2nd ed.; Mentzer, J., DeWitt, W., Keebler, J., Min, S., Nix, N., Smith, C., Zacharia, Z., Thomson Learning: London, 2001; Defining Supply Chain Management. *J. Bus. Log.* **2003,** *22* (2), 1–25.

Whipple, J. M.; Frankel, R. Strategic Alliance Success Factors. *J. Supply Chain Manag.* **2000,** *36* (3), 21–28.

Wight, O. In *The Oliver Wight ABCD Check List*, 4th ed.; John Wiley: New York, NY, 1993.

Wong, A.; Fung, P. Total Quality Management in the Construction Industry in Hong Kong: A Supply Chain Management Perspective. *Total Qual. Manag.* **1999,** *10* (2), 199–208.

Yeh, D. Y.; Cheng, C. H.; Chi, M. L. A Modified Two-Tuple FLC Model for Evaluating the Performance of SCM: By the Six Sigma DMAIC Process. *Appl. Soft Comput.* **2007,** *7*, 1027–1034.

Zhu, Q.; Sarkis, J.; Lai, K. Confirmation of a Measurement Model for Green Supply Chain Management Practices Implementation. *Int. J. Prod. Econ.* **2008,** *111*, 261–273.

Ziilal, G.; Arikan, F. Application of Fuzzy Decision-Making in Part Machining Grouping. *Int. J. Prod. Econ.* **2000,** *63*, 181–193.

APPENDIX

APPENDIX A - STRUCTURED QUESTIONNAIRE

PART A: GENERAL INFORMATION ABOUT ORGANIZATION

1. Name of Industry:
2. Location:
3. Year of Establishment:
4. Number of Employees in Organization:
5. Annual Turnover of Organization:
6. Number of Department:
7. Name of Department:
8. Product of Your Company:
9. Name of Person Contacted:
10. Department & Designation:
11. Year of Experience at Present Designation & Company:
12. Email & mobile number:

PART B: SCM & SCPM IN TEXTILE INDUSTRY

1. What is SCM?
2. What are the importance/ benefits of SCM in process industry?
3. What do textile companies think for SCM?
4. Does your company follow SCM philosophy?
   a. Yes b. No
5. What are different drivers/ key elements of SCM (Generalized)?
6. What are the different drivers of SCM in textile industry?
7. Please explain concept of SCPM in textile industry.
8. Does any SCPMS exist in your company?
   a. If yes, Please explain
   b. Name & Designation of Person handling SCPM process.
   c. If no, Why
      i. Lack of Awareness

   ii. Lack of technically skilled manpower
   iii. Never felt the need
   iv. Unable to implement as activities in industry are disorganized.

9. How performance measuring process for SCM evolving in textile industries?
10. What are the probable activities of supply chain network that can be considered for SCPM process?
    a. List them
    b. Please give your opinion (Please tick whichever is applicable);

Sr. No.	Supply chain network functional activities/ operation	Agree	Disagree
1	Planning		
2	Design		
3	Procurement		
4	Production		
5	Distribution		

11. Please suggest the potential sub-areas of supply chain network's functional activities/ operation.
12. Please comment, the use of Balanced scorecard approach for performance evaluation of supply chain using its four perspectives (Financial, Customer, Internal Business, Innovation & Learning)
13. Which type of criteria would be most preferred for SCPM? (Tick as applicable)
    a. Financial
    b. Non-financial
    c. Mixed of both

PART C: DEVELOPMENT OF CONCEPTUAL FRAMEWORK FOR SCPMS

14. Please elaborate in detail the criteria and associated sub-criteria. Rate them as per scale given below;

Sr. No.	Criteria	Strongly Disagree	Disagree	Neither	Agree	Strongly Agree
1	Financial Perspective					

2	Customer Perspective
3	Internal Business Perspective
4	Innovation & Learning Perspective
5	Environmental Perspective

16. Enlist alternatives with respect to consensus achieved in Part (B)
17. Please comment on proposal of designing hierarchical framework with
    a. Goal-Overall Supply Chain Performance Measurement using BSC approach
    b. Criteria-
    c. Sub-criteria-
    d. Alternatives-

Also, suggest that what should be the focus of this proposed research.

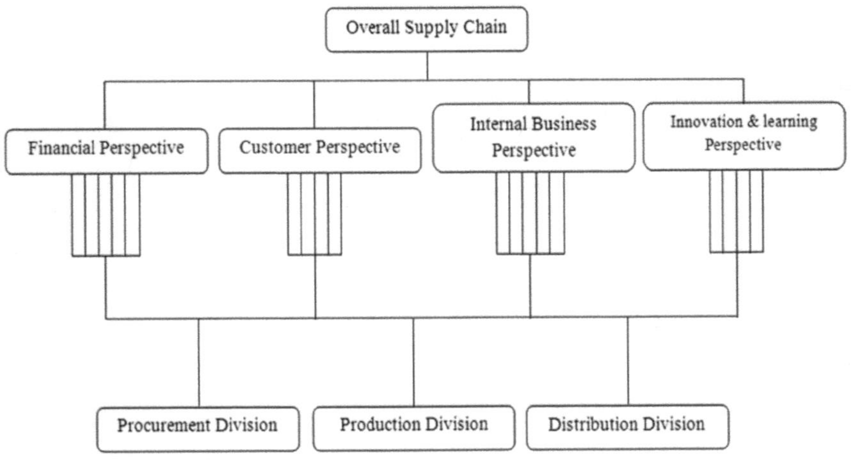

PART D: IDENTIFYING CRITICAL PERFORMANCE SUB-MEASURES FROM PROPOSED FRAMEWORK USING AHP

With respect to your early feedback and responses, you are requested to select the degree of importance on the basis of Saaty's scale for pairwise comparison of criteria, sub-criteria and alternatives.

1 – The least important
3 – Less important
5 – Moderately important
7 – More important
9 – The most important
2, 4, 6, 8- Intermediate values between the two adjacent judgments

TABLE 1 Pairwise Comparison of Alternatives w r t Overall Goal

Sr. No.	Alternatives	1	2	3
1	Procurement Division			
2	Production Division			
3	Distribution Division			

TABLE 2 Pair wise Comparison of alternatives w r t Financial Perspective

Sr. No.	Alternatives	1	2	3
1	Procurement Division			
2	Production Division			
3	Distribution Division			

TABLE 3 Pair wise Comparison of alternatives w r t Customer Perspective

Sr. No.	Alternatives	1	2	3
1	Procurement Division			
2	Production Division			
3	Distribution Division			

TABLE 4 Pair wise Comparison of alternatives w r t Internal Business Perspective

Sr. No.	Alternatives	1	2	3
1	Procurement Division			
2	Production Division			
3	Distribution Division			

TABLE 5 Pairwise Comparison of alternatives w r t Innovation & Learning Per spective

Sr. No.	Alternatives	1	2	3
1	Procurement Division			
2	Production Division			
3	Distribution Division			

TABE 6 Pair wise Comparison of criteria

Sr No	Criteria	1	2	3	4
1	Financial Perspective				
2	Customer Perspective				
3	Internal Business Perspective				
4	Innovation & Learning Perspective				

TABE 7 Pair wise Comparison of Financial Perspectives Sub-Criteria

Sr No	Sub-criteria	1	2	3	4	5	6
1	Rate of Return on Investment						
2	Supplier's Cost Saving Initiative						
3	Information Carrying Cost						
4	Inventory Carrying Cost						
5	Manufacturing Cost						
6	Cost per operation hour						

TABE 8 Pairwise Comparison of Customer Perspectives Sub-Criteria

Sr No	Sub-criteria	1	2	3	4	5
1	Quality of delivered goods					
2	Delivery performance & lead time					
3	Customer Satisfaction					
4	Effectiveness of distribution planning schedule					
5	Customer Query Time					
6	Range of Product & Services					

TABE 9 Pairwise Comparison of Internal Business Perspectives Sub-Criteria

Sr No	Sub-criteria	1	2	3	4	5	6
1	Flexibility to meet particular customer needs						
2	Total supply chain cycle time						
3	Product development cycle time						
4	Capacity Utilization						
5	Planned process cycle time						
6	Extent of cooperation to improve quality						

TABE 10 Pair wise Comparison of Internal Business Perspectives Sub-Criteria

Sr No	Sub-criteria	1	2	3	4	5
1	Buyer-supplier partnership level					
2	Accuracy of forecasting					
3	Employing IT & KM technologies					
4	Use of QE & QM techniques					
5	Employee satisfaction and skill orientation					

➢ Pair wise comparison of Alternatives w r t financial sub-criteria

A: - Return on Investment

Sr. No.	Alternative	1	2	3
1	Procurement Division			
2	Production Division			
3	Distribution Division			

B: - Supplier's Cost Saving Initiative

Sr. No.	Alternative	1	2	3
1	Procurement Division			
2	Production Division			
3	Distribution Division			

C: - Information Carrying Cost

Sr. No.	Alternative	1	2	3
1	Procurement Division			
2	Production Division			
3	Distribution Division			

D: - Inventory Carrying Cost

Sr. No.	Alternative	1	2	3
1	Procurement Division			
2	Production Division			
3	Distribution Division			

E: - Manufacturing Cost

Sr. No.	Alternative	1	2	3
1	Procurement Division			
2	Production Division			
3	Distribution Division			

F: - Cost per hour of operation

Sr. No.	Alternative	1	2	3
1	Procurement Division			
2	Production Division			
3	Distribution Division			

➢ Pairwise comparison of Alternatives w r t Customer sub-criteria

G: - Quality of goods delivered

Sr. No.	Alternative	1	2	3
1	Procurement Division			
2	Production Division			
3	Distribution Division			

H: - Delivery performance & lead time

Sr. No.	Alternative	1	2	3
1	Procurement Division			
2	Production Division			
3	Distribution Division			

I: - Customer Satisfaction

Sr. No.	Alternative	1	2	3
1	Procurement Division			
2	Production Division			
3	Distribution Division			

J: - Effectiveness to meet particular customer needs

Sr. No.	Alternative	1	2	3
1	Procurement Division			
2	Production Division			
3	Distribution Division			

K: - Customer Query Time

Sr. No.	Alternative	1	2	3
1	Procurement Division			
2	Production Division			
3	Distribution Division			

L: - Range of Product & Services

Sr. No.	Alternative	1	2	3
1	Procurement Division			
2	Production Division			
3	Distribution Division			

➢ Pair wise comparison of Alternatives w r t Internal Business sub-criteria

M: - Flexibility to meet particular customer needs

Sr. No.	Alternative	1	2	3
1	Procurement Division			
2	Production Division			
3	Distribution Division			

N: - Total supply chain cycle time

Sr. No.	Alternative	1	2	3
1	Procurement Division			
2	Production Division			
3	Distribution Division			

O: - Product development cycle time

Sr. No.	Alternative	1	2	3
1	Procurement Division			
2	Production Division			
3	Distribution Division			

P: - Capacity utilization

Sr. No.	Alternative	1	2	3
1	Procurement Division			
2	Production Division			
3	Distribution Division			

Q: - Planned process cycle time

Sr. No.	Alternative	1	2	3
1	Procurement Division			
2	Production Division			
3	Distribution Division			

R: - Extent of cooperation to improve quality

Sr. No.	Alternative	1	2	3
1	Procurement Division			
2	Production Division			
3	Distribution Division			

➢ Pair wise comparison of Alternatives w r t Innovation & Learning sub-criteria

S: - Buyer-supplier partnership level

Sr. No.	Alternative	1	2	3
1	Procurement Division			
2	Production Division			
3	Distribution Division			

T: - Accuracy of forecasting

Sr. No.	Alternative	1	2	3
1	Procurement Division			
2	Production Division			
3	Distribution Division			

U: -Employing Information Technology & Knowledge Management technologies

Sr. No.	Alternative	1	2	3
1	Procurement Division			
2	Production Division			
3	Distribution Division			

V: - Use of Quality Engineering & Quality Management techniques

Sr. No.	Alternative	1	2	3
1	Procurement Division			
2	Production Division			
3	Distribution Division			

W: - Employee Satisfaction & Skills Orientation

Sr. No.	Alternative	1	2	3
1	Procurement Division			
2	Production Division			
3	Distribution Division			

APPENDIX B: QUESTIONNAIRE

PART A: GENERAL INFORMATION ABOUT ORGANIZATION

1. Name of Industry:
2. Location:
3. Year of Establishment:
4. Number of Employees in Organization:
5. Annual Turnover of Organization:
6. Number of Department:
7. Name of Department:

8. Product of Your Company:
9. Name of Person Contacted:
10. Department & Designation:
11. Year of Experience at Present Designation & Company:
12. Email & mobile number:

PART B: SCM AND SCPM IN TEXTILE INDUSTRY

1. What is SCM?
2. What are the importance/benefits of SCM in process industry?
3. What do textile companies think for SCM?
4. Does your company follow SCM philosophy?
   a. Yes b. No
5. What are different drivers/key elements of SCM (Generalized)?
6. What are the different drivers of SCM in textile industry?
7. Please explain concept of SCPM in textile industry.
8. Does any SCPMS exist in your company?
   a. If yes, Please explain
   b. Name and Designation of Person handling SCPM process.
   c. If no, Why
      i. Lack of Awareness
      ii. Lack of technically skilled manpower
      iii. Never felt the need
      iv. Unable to implement as activities in industry are disorganized.
9. How performance measuring process for SCM evolving in textile industries?
10. What are the probable activities of supply chain network that can be considered for SCPM process?
    a. List them
    b. Please give your opinion (Please tick whichever is applicable);

Sr. No.	Supply chain network functional activities/operation	Agree	Disagree
1	Planning		
2	Design		
3	Procurement		
4	Production		
5	Distribution		

11. Please suggest the potential sub-areas of supply chain network's functional activities/operation.
12. Please comment, the use of Balanced scorecard approach for performance evaluation of supply chain using its four perspectives (Financial, Customer, Internal Business, Innovation and Learning)
13. Which type of criteria would be most preferred for SCPM? (Tick as applicable)
    a. Financial
    b. Non-financial
    c. Mixed of both

PART C: DEVELOPMENT OF CONCEPTUAL FRAMEWORK FOR SCPMS

14. Please elaborate in detail the criteria and associated sub-criteria. Rate them as per scale given below;

Sr. No.	Criteria	Strongly disagree	Disagree	Neither	Agree	Strongly agree
1	Financial perspective					
2	Customer perspective					
3	Internal business perspective					
4	Innovation and learning perspective					
5	Environmental perspective					

15. Enlist alternatives with respect to consensus achieved in Part (B)
16. Please comment on proposal of designing hierarchical framework with

a. Goal-Overall Supply Chain Performance Measurement using BSC approach
b. Criteria-
c. Sub-criteria-
d. Alternatives-

Also, suggest that what should be the focus of this proposed research.

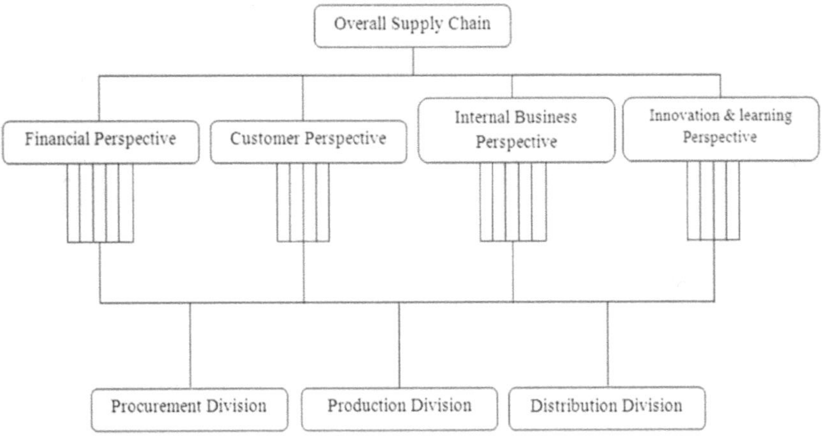

PART D: IDENTIFYING CRITICAL PERFORMANCE SUB-MEASURES FROM PROPOSED FRAMEWORK USING AHP

With respect to your early feedback and responses, you are requested to select the degree of importance on the basis of Saaty's scale for pairwise comparison of criteria, sub-criteria, and alternatives.

1 – The least important

3 – Less important

5 – Moderately important

7 – More important

9 – The most important

2, 4, 6, 8- Intermediate values between the two adjacent judgments

TABLE 1 Pairwise Comparison of Alternatives w.r.t. Overall Goal.

Sr. No.	Alternatives	1	2	3
1	Procurement division			
2	Production division			
3	Distribution division			

TABLE 2 Pairwise Comparison of Alternatives w r t Financial Perspective.

Sr. No.	Alternatives	1	2	3
1	Procurement division			
2	Production division			
3	Distribution division			

TABLE 3 Pairwise Comparison of Alternatives w r t Customer Perspective.

Sr. No.	Alternatives	1	2	3
1	Procurement division			
2	Production division			
3	Distribution division			

TABLE 4 Pairwise Comparison of Alternatives w r t Internal Business Perspective.

Sr. No.	Alternatives	1	2	3
1	Procurement division			
2	Production division			
3	Distribution division			

TABLE 5 Pairwise Comparison of Alternatives w r t Innovation and Learning Per Spective.

Sr. No.	Alternatives	1	2	3
1	Procurement division			
2	Production division			
3	Distribution division			

TABLE 6 Pairwise Comparison of Criteria.

Sr. No	Criteria	1	2	3	4
1	Financial Perspective				
2	Customer Perspective				
3	Internal Business perspective				
4	Innovation and learning perspective				

TABLE 7 Pairwise Comparison of Financial Perspectives Sub-criteria.

Sr. No	Sub-criteria	1	2	3	4	5	6
1	Rate of return on investment						
2	Supplier's cost saving initiative						
3	Information carrying cost						
4	Inventory carrying cost						
5	Manufacturing cost						
6	Cost per operation hour						

TABLE 8 Pairwise Comparison of Customer Perspectives Sub-criteria.

Sr. No	Sub-criteria	1	2	3	4	5
1	Quality of delivered goods					
2	Delivery performance and lead time					
3	Customer satisfaction					
4	Effectiveness of distribution planning schedule					
5	Customer query time					
6	Range of product and services					

TABLE 9 Pairwise Comparison of Internal Business Perspectives Sub-criteria.

Sr. No	Sub-criteria	1	2	3	4	5	6
1	Flexibility to meet particular customer needs						
2	Total supply chain cycle time						
3	Product development cycle time						
4	Capacity utilization						
5	Planned process cycle time						
6	Extent of cooperation to improve quality						

TABLE 10 Pairwise Comparison of Internal Business Perspectives Sub-criteria.

Sr. No	Sub-criteria	1	2	3	4	5
1	Buyer-supplier partnership level					
2	Accuracy of forecasting					
3	Employing IT and KM technologies					
4	Use of QE and QM techniques					
5	Employee satisfaction and skill orientation					

PAIRWISE COMPARISON OF ALTERNATIVES W R T FINANCIAL SUB-CRITERIA

A: - Return on Investment

Sr. No.	Alternative	1	2	3
1	Procurement division			
2	Production division			
3	Distribution division			

B: - Supplier's Cost Saving Initiative

Sr. No.	Alternative	1	2	3
1	Procurement division			
2	Production division			
3	Distribution division			

C: - Information Carrying Cost

Sr. No.	Alternative	1	2	3
1	Procurement division			
2	Production division			
3	Distribution division			

D: - Inventory Carrying Cost

Sr. No.	Alternative	1	2	3
1	Procurement division			
2	Production division			
3	Distribution division			

E: - Manufacturing Cost

Sr. No.	Alternative	1	2	3
1	Procurement division			
2	Production division			
3	Distribution division			

F: - Cost Per Hour of Operation

Sr. No.	Alternative	1	2	3
1	Procurement division			
2	Production division			
3	Distribution division			

PAIRWISE COMPARISON OF ALTERNATIVES W R T CUSTOMER SUB-CRITERIA

G: - Quality of Goods Delivered

Sr. No.	Alternative	1	2	3
1	Procurement division			
2	Production division			
3	Distribution division			

H: - Delivery Performance and Lead Time

Sr. No.	Alternative	1	2	3
1	Procurement division			
2	Production division			
3	Distribution division			

I: - Customer Satisfaction

Sr. No.	Alternative	1	2	3
1	Procurement division			
2	Production division			
3	Distribution division			

J: - Effectiveness to Meet Particular Customer Needs

Sr. No.	Alternative	1	2	3
1	Procurement division			
2	Production division			
3	Distribution division			

K: - Customer Query Time

Sr. No.	Alternative	1	2	3
1	Procurement division			
2	Production division			
3	Distribution division			

L: - Range of Product and Services

Sr. No.	Alternative	1	2	3
1	Procurement division			
2	Production division			
3	Distribution division			

PAIRWISE COMPARISON OF ALTERNATIVES W R T INTERNAL BUSINESS SUB-CRITERIA

M: - Flexibility to Meet Particular Customer Needs

Sr. No.	Alternative	1	2	3
1	Procurement division			
2	Production division			
3	Distribution division			

N: - Total Supply Chain Cycle Time

Sr. No.	Alternative	1	2	3
1	Procurement division			
2	Production division			
3	Distribution division			

O: - Product Development Cycle Time

Sr. No.	Alternative	1	2	3
1	Procurement Division			
2	Production Division			
3	Distribution Division			

P: - Capacity Utilization

Sr. No.	Alternative	1	2	3
1	Procurement division			
2	Production division			
3	Distribution division			

Q: - Planned Process Cycle Time

Sr. No.	Alternative	1	2	3
1	Procurement division			
2	Production division			
3	Distribution division			

R: - Extent of Cooperation to Improve Quality

Sr. No.	Alternative	1	2	3
1	Procurement division			
2	Production division			
3	Distribution division			

PAIRWISE COMPARISON OF ALTERNATIVES W R T INNOVATION AND LEARNING SUB-CRITERIA

S: - Buyer-Supplier Partnership Level

Sr. No.	Alternative	1	2	3
1	Procurement division			
2	Production division			
3	Distribution division			

T: - Accuracy of Forecasting				
Sr. No.	**Alternative**	**1**	**2**	**3**
1	Procurement division			
2	Production division			
3	Distribution division			

U: - Employing Information Technology and Knowledge Management Technologies

Sr. No.	Alternative	1	2	3
1	Procurement division			
2	Production division			
3	Distribution division			

V: - Use of Quality Engineering and Quality Management Techniques

Sr. No.	Alternative	1	2	3
1	Procurement division			
2	Production division			
3	Distribution division			

W: - Employee Satisfaction and Skills Orientation

Sr. No.	Alternative	1	2	3
1	Procurement division			
2	Production division			
3	Distribution division			

Appendix C- Priorities of Supply Chain Cyclic Processes with Respect to Performance Sub-criteria for Case Company 1.

	Financial Perspective						Customer Perspective					
Level of Sub-criteria	St	Tt	Op	Op	Op	Op	Op	St	S, T, O	Tt	St	St
Supply Chain Cyclic Processes	A	B	C	D	E	F	G	H	I	J	K	L
Procurement	0.217	0.461	0.406	0.201	0.272	0.227	0.196	0.246	0.266	0.275	0.331	0.27
Production	0.498	0.319	0.328	0.519	0.483	0.528	0.539	0.461	0.405	0.261	0.285	0.57
Distribution	0.285	0.221	0.266	0.279	0.245	0.245	0.265	0.293	0.329	0.464	0.384	0.16
Max. Eigen Value (λ max)	3.016	3.006	3.099	3.035	3.005	3.002	3.098	3.001	3.012	3.002	3.018	3.01
Consistency Index	0.002	0.003	0.048	0.017	0.0025	0.001	0.049	0.0005	0.006	0.001	0.009	0.005
Consistency Ratio	0.0034	0.0051	0.08	0.03	0.0043	0.0017	0.08	0.0008	0.01	0.0017	0.015	0.008

	Internal Business Process Perspective						Innovation & Learning Perspective				
Level of Sub-criteria	St	St	Tt	Op	Tt	S, T, O	St, Tt	St	S, T, O	S, T, O	S, T, O
Supply Chain Cyclic Processes	M	N	O	P	Q	R	S	T	U	V	W
Procurement	0.263	0.356	0.265	0.186	0.322	0.25	0.488	0.456	0.301	0.226	0.283
Production	0.554	0.444	0.526	0.507	0.438	0.575	0.253	0.182	0.462	0.603	0.508
Distribution	0.184	0.2	0.21	0.307	0.238	0.175	0.258	0.362	0.253	0.169	0.209
Max. Eigen Value (λ max)	3.0261	3.011	3.001	3.0025	3.0136	3.0082	3.001	3.0189	3.001	3.0024	3.003
Consistency Index	0.013	0.005	0.0005	0.00125	0.0068	0.0041	0.005	0.0094	0.0005	0.0012	0.0015
Consistency Ratio	0.022	0.0094	0.00086	0.0021	0.0117	0.007	0.0008	0.016	0.00086	0.00206	0.0025

Appendix D - Priorities of Supply Chain Cyclic Processes with Respect to Performance Sub-criteria for Case Company 2.

	Financial Perspective						Customer Perspective						Internal Business Process Perspective						Innovation & Learning Perspective				
Level of Sub-criteria	St	Tt	Op	Op	Op	Op	Op	St	S, T, O	Tt	St	St	St	St	Tt	Op	Tt	S, T, O	St, Tt	St	S, T, O	S, T, O	S, T, O
Supply Chain Cyclic Processes	A	B	C	D	E	F	G	H	I	J	K	L	M	N	O	P	Q	R	S	T	U	V	W
Procurement	0.285	**0.462**	**0.378**	0.33	**0.377**	0.347	**0.381**	0.332	**0.357**	0.347	0.352	**0.426**	0.365	**0.408**	0.35	0.266	**0.382**	**0.403**	**0.439**	**0.419**	**0.425**	0.306	**0.363**
Production	**0.384**	0.236	0.312	**0.385**	0.331	**0.403**	0.369	**0.358**	0.352	0.249	0.286	0.325	**0.387**	0.311	**0.382**	**0.405**	0.35	0.35	0.238	0.225	0.313	**0.407**	0.356
Distribution	0.331	0.302	0.307	0.285	0.289	0.249	0.25	0.309	0.29	**0.403**	**0.362**	0.248	0.25	0.281	0.267	0.329	0.267	0.247	0.323	0.357	0.262	0.287	0.281
Max. Eigen Value (λ max)	3.003	3.001	3.002	3.004	3.003	3.006	3.005	3.0035	3.0022	3.0018	3.0022	3.001	3.002	3.001	3.0013	3.0004	3.001	3.0006	3.002	3.005	3.003	3.001	3.0011
Consistency Index	0.001	0.0005	0.001	0.002	0.001	0.003	0.0025	0.00175	0.0011	0.0009	0.0011	0.0005	0.001	0.0005	0.00065	0.0002	0.0005	0.0003	0.001	0.0025	0.0015	0.0005	0.00055
Consistency Ratio	0.0025	0.0008	0.0017	0.003	0.0025	0.005	0.0043	0.00301	0.00189	0.0015	0.00189	0.00086	0.0017	0.00086	0.0011	0.00034	0.00086	0.00051	0.0017	0.0043	0.0025	0.00086	0.00094

Appendix E - Priorities of Supply Chain Cyclic Processes with Respect to Performance Sub-criteria for Case Company 3.

	Financial Perspective						Customer Perspective						Internal Business Process Perspective						Innovation & Learning Perspective				
Level of Sub-criteria	St	Tt	Op	Op	Op	Op	Op	St	S, T, O	Tt	St	St	St	St	Tt	Op	Tt	S, T, O	St, Tt	St	S, T, O	S, T, O	S, T, O
Supply Chain Cyclic Processes	A	B	C	D	E	F	G	H	I	J	K	L	M	N	O	P	Q	R	S	T	U	V	W
Procurement	0.239	0.361	0.309	0.243	0.228	0.225	0.263	0.233	0.303	0.322	0.329	0.292	0.288	0.368	0.293	0.181	0.344	0.344	0.407	0.428	0.304	0.26	0.285
Production	0.471	0.281	0.454	0.501	0.499	0.542	0.444	0.486	0.428	0.234	0.289	0.463	0.466	0.351	0.444	0.48	0.389	0.389	0.233	0.194	0.446	0.448	0.382
Distribution	0.29	0.358	0.237	0.255	0.273	0.233	0.293	0.281	0.269	0.444	0.382	0.245	0.246	0.281	263	0.34	0.267	0.267	0.36	0.377	0.249	0.292	0.333
Max. Eigen Value (λ max)	3.008	3.0008	3.004	3.004	3.1	3.002	3.006	3.004	3.008	3.0132	3.016	3.1	3.02	3.06	3.0018	3.006	3.0016	3.016	3.014	3.04	3.001	3.0006	3.008
Consistency Index	0.004	0.0004	0.002	0.002	0.05	0.001	0.003	0.002	0.004	0.0066	0.008	0.05	0.01	0.03	0.0009	0.003	0.0008	0.008	0.007	0.02	0.0005	0.0003	0.004
Consistency Ratio	0.006	0.0006	0.0034	0.0034	0.086	0.0017	0.005	0.0034	0.0068	0.011	0.013	0.086	0.017	0.051	0.0016	0.0051	0.0013	0.013	0.012	0.034	0.0009	0.00051	0.0068

Appendix F- Priorities of Supply Chain Cyclic Processes with Respect to Performance Sub-criteria for Case Company 4.

	Financial Perspective						Customer Perspective						Internal Business Process Perspective						Innovation & Learning Perspective				
Level of Sub-criteria	St	Tt	Op	Op	Op	Op	Op	St	S, T, O	Tt	St	St	St	St	Tt	Op	Tt	S, T, O	St, Tt	St	S, T, O	S, T, O	S, T, O
Supply Chain Cyclic Processes	A	B	C	D	E	F	G	H	I	J	K	L	M	N	O	P	Q	R	S	T	U	V	W
Procurement	0.216	0.435	0.409	0.256	0.252	0.288	0.257	0.221	0.219	0.303	0.329	0.287	0.32	0.307	0.263	0.229	0.361	0.26	0.439	0.46	0.303	0.248	0.256
Production	0.496	0.219	0.359	0.483	0.529	0.523	0.487	0.517	0.484	0.269	0.269	0.511	0.46	0.429	0.526	0.487	0.352	0.527	0.238	0.221	0.479	0.511	0.464
Distribution	0.288	0.345	0.233	0.261	0.219	0.19	0.256	0.262	0.297	0.429	0.402	0.202	0.221	0.264	0.211	0.284	0.287	0.214	0.323	0.319	0.218	0.241	0.28
Max. Eigen Value (λ max)	3.0004	3.1	3.0016	3.0011	3.0036	3.0178	3.005	3.0004	3.018	3.0074	3.0075	3.04	3.0032	3.0098	3.0004	3.08	3.1	3.08	3.0128	3.0028	3.0094	3.08	3.018
Consistency Index	0.0002	0.05	0.0008	0.0055	0.0018	0.0089	0.0025	0.0002	0.009	0.0037	0.0038	0.02	0.0016	0.0049	0.0002	0.04	0.05	0.04	0.0064	0.00148	0.0047	0.04	0.009
Consistency Ratio	0.0003	0.086	0.0013	0.0094	0.0031	0.015	0.0043	0.00034	0.015	0.0063	0.0064	0.034	0.0027	0.0084	0.00034	0.068	0.086	0.068	0.011	0.0024	0.0081	0.068	0.015

INDEX

J

K

M

O

P

Q

R

S

T

U

W